# The Comfort Trilogy

## The Prisoners of Comfort

## The Rebels of Comfort

## The Dangerous Book for Pharmacists

By: Jim Plagakis, R.Ph.

Coyright by Jim Plagakis 2011

**Your ISBN:** 978-1-257-76132

For twenty years Jim Plagakis has been the author of the popular Drug Topics Magazine column JP at Large. He has been a prescient observer of the drug store industry and his predictions have been consistently accurate. Jim loves pharmacy and he knows that it is the job that can present problems. The profession is just fine.

Mister Plagakis was one of the first to identify institutionalization of the pharmacist in the modern chain drug store culture. This social phenomenon is rather recent and became pronounced beginning in the early 1990s when the industry began depending on computer programs for productivity. Reports became the rulers of the pharmacist's behavior and prescription counts became the dictator of their time. Moore's Law regards exponential increases. Pharmacists filled around $30 Billion worth of Rx in 1989. $105 billion worth of Rx in 1999. $234 billion in 2008. How long will it take to get to a trillion dollars? Moore's Law says not very long. It is no wonder that keeping the Prescription Mills grinding is so important to the big drug store chains. The price is that highly educated medical professionals are asked to dumb down and toe the mark.

*"There must be some way out of here," said the joker to the thief*
*"There's too much confusion, I can't get no relief*
*Businessmen, they drink my wine, plowmen dig my earth*
*None of them along the line know what any of it is worth"*

*"No reason to get excited," the thief, he kindly spoke*
*"There are many here among us who feel that life is but a joke*
*But you and I, we've been through that, and this is not our fate*
*So let us not talk falsely now, the hour is getting late"*

*Bob Dylan*

"In a hole in the ground there lived a hobbit. Not a nasty, dirty, wet hole, filled with the ends of worms and an oozy smell, nor yet a dry, bare, sandy hole with nothing in it to sit down on or to eat: It was a hobbit-hole, and that means comfort."

John Ronald Reuel Tolkien

# The Prisoners of Comfort

## Institutionalization

To put someone into an institution such as a prison. To make something an established custom or accepted part of the (job) A good example is simply going to the bathroom when you have to go. Pharmacists will consistently answer one more phone call, check one more prescription and answer one more question. Often, they make it to the bathroom at the crisis stage and end up with damp underwear.

## The Players

### Pharmacists at The Prisoners

### The Non-Pharmacist Managers are the Jailers

### The Middle-Managers (District) are the Wardens

### The Executives are the Governors

# Pharmacy Culture is: The Prison

I am an institutionalized pharmacist. I have been this way so long that it is part of who I am. I am used to eating my meals at work standing up. I am used to being interrupted by the telephone, a patient with a question or any of a number of other reasons that my meal gets cold if I am stupid enough that day to bring something that is better warm rather than my usual cold sandwich. I drink too many caffeinated beverages and I succumb to Snickers bars, salty chips and cookies. In the past 43 years, I may have actually taken a real lunch, when I could leave the store, a couple hundred times. That's only six times a year. It did not take me very long in 1965 to get with the program and surrender to the established lunch time custom in pharmacy. Eating on the run.

My legs are shot. I have worked so many twelve and thirteen hour days without a break that it is amazing that I can still stand for eight hours. I am so used to standing all day long that I even stand when it is not busy and the stool is right beside me. I actually feel uncomfortable sitting down at work. I have Post-Polio Syndrome and concomitant neuropathy in my legs. I can't blame the pharmacy culture for that, but I have not used good judgment. Pharmacists stand and they stand all day long. That's the culture so that is what I did and I am still doing it forty three years after the starting gun went off.

Pharmacies are notoriously poorly designed for the health of pharmacists. The computer screens are fixed at a certain height. What is appropriate for a man six foot two is not right for a girl five foot four. The floors at too many places have no comfort pads or cheap imitations. Holding the telephone between your ear and shoulder can cause painful conditions in your neck. In January, 2010, I turned myself over to two Physical Therapists. I had terrible neck and shoulder pain from a head-forward posture after four decades of bending over the counter. It took almost a year, but I have followed the exercise program religiously and stand straight again, like when I was a boy. Why did I put up with it for 40 years?

I believe that pharmacists should be managed by pharmacists. I do not believe that it is appropriate for a highly educated medical professional to be managed by a non-pharmacist manager who may or may not have any education beyond high school. I know that that sounds elitist, but so be it. Non-Pharmacist Managers are the Jailers in our pharmacy culture. They are more often than not products of the culture. For the Non-Pharmacist Manager, company rules trump the pharmacist's personal ethics, professional standards and legal responsibilities. The Jailers enforce the company's rules and the mandates for completing reports on time. The Jailers are particularly zealous about the Prescription Mill. Some companies have timers that report how fast the pharmacist runs The Prescription Mill via computer-generated reports. If you lag behind in the culture of some Chain Drug Stores, you are brought onto the carpet, even written up.

There have been some Non-Pharmacist Managers who did not like me. They did not like that I did not do what they wanted me to do just because they were ordering me to do it. If it was a professional matter, the Non-Pharmacist Manager had absolutely no standing. If it was a pharmacy business matter, I would consistently make my own choices or confer with The Warden, the Pharmacist District Manager. A few Jailers absolutely hated me. They hated that I did not work as many hours as they did and that I earned more money than they did. They were unhappy managers and that made them even more aggressive about making me abide by the rules of The Prison. Sometimes I did and sometimes I did not. I was one of those pharmacists who the Jailers wanted in "the hole". A disclaimer. I have worked with Non-Pharmacist Managers who were happy Jailers. They recognized that I was a talented Pharmacy Manager and that what I did made their bonus checks fatter.

Store managers can be petty, mean and spiteful. They look for every chance to criticize your behavior. They love to write the Prisoners up. I had a non-pharmacist manager who insisted that I sign up on the calendar for bathroom cleaning detail. I was a janitor for awhile in college so I know how to clean. When my day came, I took the vacuum, mop and pail, electric buffer, Windex, sink cleaner and other assorted supplies into the men's room. I was having a merry time cleaning. It was a terrific break from The Prescription Mill. After only twenty minutes, the store manager burst into the men's room. He demanded to know what I was doing. I reminded him of the cleaning schedule. He was so angry there was a spray of spit as he shouted at me, demanding that I get back to the pharmacy. Another manager wrote me up because I refused to wear the black slacks and white shirts that were company policy. I reminded him that Washington State law required the company to purchase any clothes that were required for the job. I had a manager tell me that I couldn't chew my food in sight of the patients. Another criticized me because I have a habit of complimenting women when they look good. He said it could be construed as sexual harassment. The Jailers will always look for weaknesses to exploit. I learned to either never show weakness or to stand my ground. I am a Prisoner, but I am a medical professional Prisoner.

Vacations are one of the benefits that the companies ballyhoo when you are first looking at the job. Most drug store chains promise multiple weeks of vacation after only one year of service. I don't think that I have enjoyed vacations in the summer more than a half dozen times. Even after years of service, they made it very difficult to plan a vacation when the kids were out of school. The Jailers often had a little grin when they told me, "Plagakis, you didn't get those two weeks in July that you wanted. I told them to put you down for one week in August and the other in November."

One week in 1976, I worked sick. There was a rotovirus in the community and every employee of the little drug store I managed was out sick with diarrhea and nausea and vomiting. I was the manager. My Saturday relief was not available during the week so I worked sick. I had to make a sign that read "Be back in a minute". I had to use the bathroom often for a couple days. I would lock the front door and put up the sign. I have worked sick too many times because that is what the culture expects from a pharmacist. The store managers figured this out quickly. They always expected me to work while I was ill.

The word "Rude" is a buzzword for The Jailers and The Wardens. Smart rat drug store customers know that if they use that word the Jailer will be all over the Prisoner. I have been accused of being rude countless times and almost every time, I have been put in the corner with the dunce cap on my head. I have been written up because customers have accused me of being rude. The Jailers seemed to relish the moments in the office when I was obliged to give my side of the story. Occasionally, I was written up. The Jailer was always furious when I refused to sign the write up form. Prisoners aren't allowed to do that. I still did. My manner may be abrupt and business-like, but I am never rude.

The Jailer will chortle with glee when he sees The Prisoner working in the pharmacy after it has closed. The order needs to be placed. The order needs to be put up. At closing, there were still thirty prescriptions to complete. Working off-the-clock is illegal in most states if you are an hourly employee. If you are a salaried employee and your pay is based on a forty hour week and you actually work sixty hours, most state laws require that you be paid. If a pharmacist is in the habit of working off-the-clock, they must document every single hour, including the reason why they stayed late or came in early. If the state doesn't have progressive laws, the federal government does.

The Prisoner will often have to work without enough help. There will be times when The Prisoner is in a situation where he will have to run The Prescription Mill, answer the telephone, attend to the Drive Through and be friendly and forthcoming at the main counter. The law as well as personal standards and professional ethics require that he counsel patients on new prescriptions. I have been in that situation numerous times. I have closed the Drive Through with the excuse that there is an electrical problem. When the manager finds out, he lambasts me. I listen carefully like a good Prisoner, but I do not promise to never do it again. The telephone is a last priority. I just do not answer it. I bounce back and forth between The Prescription Mill and the main counter. I do my best and my number one rule is everyone has to wait. The Governors are at fault when there is not enough help. The Governors care only about numbers. Primarily the bottom line. They are the ones who sign contracts with the PBMs that put the company at a disadvantage right out of the starting gate. The profits are pitifully thin. The Governors then go to the Wardens and demand to see better results. The Wardens then go where they always go. They go to The Jailers and demand that the pharmacy payroll be pared down. It is always technicians who lose hours and The Prisoners have to work without enough help. Every pharmacist knows that it is customer service that is affected negatively. The Jailer doesn't care. It is just one more chance to blame the pharmacists for something that is not their fault.

Just about every pharmacist has professional standards that are not being fulfilled. It bothers them that they keep their noses in The Prescription Mill and that they do not live up to their obligations to provide superior patient care by counseling when appropriate. It bothers them that they do not have the time to give personal attention to patients who need assistance with OTC items. They know that OTC products are drugs and just as dangerous as Rx Only products.

Some chain drug stores have timers. At least one chain is driven by the timers. The Prisoner is on a "chain gang". Unfortunately for the profession, there are pharmacists who have given up. They view their situation as lifers with no possibility of parole. They accept the shackles. They trudge along. They are happy to collect their handsome paychecks, but make no attempt to make the choices that could change their lives. Worst of all, they blame the profession.

Being a Prisoner is not a position in which a person can thrive. There is little that a pharmacist can do to become a self-actualized person. The Prisoner's life is a set of rules. Some Prisoners rebel and overreact aggressively and quit or get fired. I quit a management job in anger once. I had worked for the company for over fifteen years. My commute was ten minutes when the traffic was bad. I felt that I was right in my dispute with a store manager. I was angry and indignant and this is a very dangerous stew. They may use food as a way out and become food addicts. They can engage in drinking too much or risky sexual habits that are often with a partner who is not a spouse. A female pharmacist was treating a vaginal infection with Metronidazole. She was an attractive petite blonde. I will always remember her perfume. She wore it tastefully, Pleasures by Elizabeth Arden. I called her a friend in arms, a sister pharmacist. She was a church-going woman with three children. She could not imagine how she was infected. I suggested that her husband could have been the culprit and she confessed that she and her husband no longer had sex. A few weeks later, the infection was back. Again, she was clueless until I asked her if she had a boyfriend. It was too much for her. She got weepy and asked me why I didn't mind my own business. I lost a friend. Soon after that, she changed jobs, then she changed towns, then she changed families. I didn't keep track. I don't even know her Christmas card address. The last I knew for sure was that she was living alone, without her three children.

Drug abuse is much more prevalent than you would think. A 2001 Drug Topics Magazine published a study by Dean Dabney, Ph.D. Georgia State University, Criminal Justice Department. 45% of Pharmacists have diverted potentially addictive pharmaceutical drugs for their own use. Near seven o'clock when the pharmacist is all alone and there are cars at the drive-through, customers standing in line at the register, six prescriptions to be filled and the phone ringing, it can be tempting to use a readily available chemical to help in getting through the day.

Other Prisoners get depressed and are barely able to function. There is a delightful artist in Seattle who probably could not survive if she identified herself as a pharmacist. For her, the Prison is so tyrannical and unfair, that her friends do not even know what she does for her living. She works a part time schedule in a chain drug store. She collects her paycheck. Pharmacy is in an armored compartment of her mind that she rarely visits.

Some pharmacists manage to remain brave, dignified and unselfish, but The Prison is not a place for courageous behavior. For many pharmacists, the conditions of the pharmacy are so oppressive that they give up. The mental conflicts are much too stressful. The clashes of will-power that they experience when they stand up to The Jailer are soul-destroying and they shrink into a true Prisoner's existence. An existence that does not even resemble the life of a highly trained medical professional.

When you are thirty years old and earn ten thousand dollars a month, you realize that your family does not have to wait for the good life. You can have it now. Making payments on your student loan at a thousand a month is not a problem. You have two nice cars, your children go to a private school, you can afford the best dance classes for your daughter and your son is already a black belt in Taekwondo. Your house is a dream that you and your spouse believed would come someday. Why wait? You told your spouse that you could have it now. The mortgage is hefty, but you can make the payments and your sisters are envious that you got that place on the lake when they may always be stuck in town. You have a life of comfort. You like it and you don't ever want to have money be a problem again. You and your spouse scrimped during the college days and you have vowed to never do it again. You may not know it, but you are a Prisoner of Comfort.

The Governors and The Wardens depend on that. They need for you to be a Prisoner. As long as Comfort is King, you may always be The Prisoner and the Pharmacy will always be your Prison.

I worked with a pharmacist in Washington State for whom there would never be enough. This man was a Prisoner of the highest order. He worked a full time job for a chain drug store and he worked weekends for a major grocery chain. He made a lot of money. He had nice things and lived in a condominium in the Cascade Mountains foothills. There was a hillside outside his living room. He could sit on his deck and hear nothing but the wind and the birds. His problem was that he didn't visit his deck very often because he was in Prison. He said that it didn't matter that his working day was boring and repetitive and that it drained his energy worse than if he had been on an actual chain gang. He declined our dinner invitation so many times that I stopped inviting him. The girls we invited to meet him were nonplussed. We said it wasn't them. He was comfortably not engaged in life. A self-imposed Stockholm Syndrome existence.

When you are in Prison, there are forces beyond your control that can take away your freedom. That is what The Prisoner believes, however you can always choose what you feel and what you do about what happens to you. You have a choice to remain brave, dignified and unselfish, or become no more than a robot.

I worked with a pharmacist in the late 1960s. We were the same age. We were in our late twenties. It was in a union shop. We took divergent paths. I was not satisfied with the everyday sameness of the job. I wanted to exercise my management muscles so I quit the union job and began a life of looking for more. I never did find enough. My friend, on the other hand, was content. He stuck with the union job until he retired. He was a dignified man and his self-respect was not dependent on the opinion of others. A union job is positively different, however. There are protections in place that make running a pharmacy like a prison very difficult. When you complain to the union, things get done.

A pharmacist's wage allows for pleasure in life. We make enough money to do the things that bring enjoyment to our everyday existence. There is some gratification in eating good food. The little bar has only top shelf libations. Our family is safe and protected in the home our wage has provided. We wear nice clothes and have all of the amenities of the 21st Century. We may belong to social organizations and we participate in amusing recreation. For most of us, this is probably not enough. Someone did say, "You can't buy happiness".

Right out of the pot, I wanted some power. I believed that a management career would allow me to exact a revolution of sorts. It all came down to money, of course. When I produced a gross profit better than any other manager in the company, I was rewarded with a transfer from California to Whidbey Island, north of Seattle. I relished in that kind of power for about fifteen minutes. I had the terrific place to raise the kids, but I found out very quickly that, in Washington State, I was more a prisoner than ever before. I now know that the perception that one has power as a pharmacy manager is false. I actually managed the pharmacy in California. I made decisions that affected the profitability of the department. I had no latitude in Oak Harbor to do anything but let the computer make the decisions. I was suddenly and completely institutionalized. I was powerless.

I want my life to mean something. My need for pleasure is simple. There is nothing better than a cup of coffee and the company of my wife or a good friend. Of course, I like it better when I am outside in a beautiful place, preferably beside a body of water. The most exquisite pleasures I remember from years ago are always like that. Simple and uncomplicated.

Power is a vaporous thing. You can observe yourself as a manager and say, "You are one powerful, Dude, man. You make all of the decisions and everybody jumps." It's still all about money. You will find out just how powerful you are when you are consistently over the payroll budget, have an overstocked pharmacy and the wait times are too long.

The question I have asked myself is: How do I insist that the job means more than the money? I found it about ten years ago. When I am getting ready for work, I say, "You will make a contribution to another human being today, Jim." And I do. We are fortunate that all of the legal ducks are lined up in our favor. We are mandated by law to counsel. The stage is set and we are the star players. We get opportunities every single day to make a difference. If we don't take those opportunities, shame on us. The Jailers have no say on this, if they are smart. This is the best, and possibly last, good chance out of The Prison. Pharmacists complain that they do not have the time. They say that they could be penalized by the Wardens if they do not make the numbers. These people are so thoroughly institutionalized that they can't see that they are only half a pharmacist, if that. There is a right way and a wrong way to practice pharmacy. In my view, a patient-centric practice is the only way.

I can't expect my job to give me meaning. I have to find meaning in my job, all by myself.

Most of us Prisoners live a very shallow existence because we are 21st Century Americans and the place where most Americans look for meaning is their job. It is classic disappointment when, early on, we realize that we are expected to pay more attention to the Prescription Mill than to the needs of our patients. As interns, we see our preceptors growing older with no spark of life in their eyes. They just want to get through the day, make the numbers and get out of there as fast as they can. The hearts are ripped out of young pharmacists when they become conscious of what they have signed up for. The Prison, with the help of the preceptors, turns young people with dreams and goals into Prisoners, ripe with energy, who can run The Mill faster than the last guy.

Happiness in a job can't come about from pure determination and hard work. Happiness and success happen as a result of dedication to a cause greater than oneself. This does not mean that your cause has to be huge, unwieldy and requiring enormous time and effort. It just means that there has to be a reason to go to work that is bigger than making a living. You cannot will success. It just happens. It comes from a well of intention that you can replenish every single day, if you want.

How does everyday work life in the pharmacy affect the dignity, self-respect and integrity of the practicing pharmacist? This is how one pharmacist handled his well-being. He was a thirty-something family man when I knew him well. His wife was an accountant who did taxes for small corporations. His little boy was given every advantage. He had security and a bright future as a pharmacy manager. He enjoyed sporting events and played tennis on his days off. He and his wife had a date once a month. Often, they would go out of town for a honeymoon-style weekend. He was still a Prisoner and there was something lacking. It was a toxic stew for a man with high personal standards. He confessed to me years later that he just wanted to be someone's hero. For him, working as a Prisoner at a Prison made him ineligible for heroism, even for his little boy.

My friend gave it all up for his ego. First, he started to drink more. That made it worse. He suffered from sleep disturbances and confessed that there was nothing in his life that brought him pleasure. During a philosophical evening powered by Pinot Noir for me and Tequila shooters for him, he claimed that he did not see any reason to continue. That bothered me. Prozac was a relatively new drug and I urged him to talk with his doctor. Instead, he went out and bought a $30,000.00 Harley Davidson.

His wife fought him. She refused to ride with him so he went and found a girl who would. He grew a biker's beard and let his hair grow. The bike and his new friends became his life. He made a joke once about swiping Vicodin, but never said a never another word about it. When I asked him, he just laughed it off. I contacted him a few months ago. It was twenty years later. This fifty-something man is now single. His twenty-something son avoids him. His ex-wife told him that she hated him. She repeated it three times before she slammed the phone. He hasn't talked to her since. He practices exclusively as a relief pharmacist and told me that the places where he works know exactly what they are getting. A bearded biker pharmacist who tolerates nothing. As he said it, "I take no shit."

I asked him, "So pharmacy for you now is just about the money?"

There was a pause and I remembered him as I saw him twenty years ago. Still handsome with a rugged look with the beard and hair.

His eyes were bright and he looked like life was being good. His voice sounded tired. I wondered how he looked now. He said, "Jamie, me boy, it has always been just about the money."

My friend was never a common pharmacist. He was given special privileges because he was a Superstar in the company and they had him pegged for middle management one day. He was never the Prisoner who would allow the Jailer to abuse him. His Jailers learned to just leave him alone. However, he made the same sacrifices that you and I have made. He suffered under the same poor working conditions. I don't know what really caused him to bolt, but my guess is that there was no "Why" good enough to make the "How" worthwhile.

There is a woman in East Cleveland with a "Why" so good that she can put up with just about any "How". Her job is her life, but it not about the money and it never was. She is dedicated to serving a clientele that is very poor. Most of them are on welfare. They are not very well educated, but this fit, blonde white woman is their angel. The Jailers are good to her. They never bother her at all and give her anything she asks for. The Wardens of her company have given The Jailers their marching orders. This woman provides the only stability in the entire store. She has no pharmacist partner. The hours she is off are covered by floaters. I have a great deal of respect for her. Her job is not to sell drugs, but to serve the patients. I have more respect for her because she takes such good care of herself. The company is pleased to give her the paid three vacations of three weeks every year that she negotiated for. She likes Cancun and the islands of Hawaii. She spent one vacation on a train, traveling across Canada.

The real difference between these two pharmacists is that one was a Prisoner with no "Why" and the other is able to thrive doing a difficult "How" because she has a worthwhile "Why". The woman in Cleveland has a steady boyfriend. He is a doctor who volunteers at the free clinic where they met.

I don't believe for a minute that you have to have a "Why" as heroic as hers is. It is her choice. She likes her life. We all have choices every day. My "Why" is much simpler and it works for me.

We go through three phases in our job of working as a pharmacist. They are distinct and, depending where you are in your career, you will recognize these important chapters.

Young pharmacists do not know that they are Prisoners when they enter the profession. They are all Doctors now. They have been clinically trained. They suffer from a psychological condition called a delusion of hope. They see how older pharmacists are treated. They watch their preceptors trudge through twelve hour shifts. They hear the staff pharmacist arguing on the telephone with their spouse. They can't make it to their child's school event again and it is not their fault. They see that the pharmacist is either too skinny or too fat from either avoiding bad food or succumbing to candy bars and caffeinated beverages with a yellow mouth from that giant bag of Cheetos. The youngsters will not hesitate to make eye contact and tell you it is not going to happen to them. It will be different where they work.

They will also tell you that they will practice pharmacy as a professional and not just run the Prescription Mill. They know exactly what they are supposed to be doing and they, by god, are going to do it right.

They have high expectations and have those prospects dashed right out of the gate. The Jailers assert their brand of control as soon as they can. The "Why" of a young pharmacist is simply to practice pharmacy in the manner in which they were trained. This is not allowed because of the demands that the Prescription Mill places on their time and energy. They deny it in the beginning, but the pressure of a waiting and impatient public get to them the first week. When the demands of a ringing telephone, people waiting at the counter, a drive-through bell ringing, doctor calls on the voice mail, twelve prescriptions still not filled and the manager commanding that they help the woman wanting vitamins defeat them and they compromise their integrity the first time, the flood gates open and they will drop counseling and professionalism more often as their career continues. They become everything they made fun of while in pharmacy school. They are a deep well of drug therapy information and they use none of it. They hate themselves.

I worked with a young woman in Washington State when she was an intern. She told me on numerous occasions that she knew that she would have to own her own store if she wanted to be happy working in pharmacy. She had watched the pharmacists she had worked with dance The Prisoners' dance and decided that she would have none of it.

Young people are, by nature, idealists. They are going to change the world. They are not going to compromise. They have been thoroughly indoctrinated by their professors that they are important members of the medical community and that their clinical skills will be utilized every day. That is the way it should be. It would be unconscionable if a professor told them that they will be chained to The Prescription Mill eight hours to fourteen hours straight, every single day.

There are many students who need to work and the well of jobs is the deepest in retail. Most of them work for one of the chains. They watch the veteran pharmacists and come to one of two possible conclusions. They say, "I'm better than John and Mildred." They have hope. "I am not going to put up with what they tolerate. I am going to practice pharmacy the way I have been taught." Or they say, "This is impossible." For these young people it is hopeless before they even start. "Look at John and Mildred. I could not tolerate being like them."

I believe that most new pharmacists enter retail believing that they are different. They keep the delusion that they will not succumb to the repressive pressures of The Prison right up until they can't anymore. They are young and very strong and they have not been pounded on yet so they walk tall for awhile until they start compromising the three pillars of a professional person on a regular basis. I am referring to personal standards, ethical principles and professional responsibilities.

Of these three, it is especially damaging to one's psyche when personal standards are thrown under the bus in order to do the "How" that is required to run The Prescription Mill. I hate it when I talk to a young pharmacist and see that there is no spark in their eyes anymore.

There is no more life there than there is in the eyes of a piece-work laborer in a factory. It isn't because something has been taken away from them. It is because they have voluntarily given away something that was personally very important.

I knew a young woman in Vermont who always said that she wanted to specialize in women's health. She didn't want to just attend to women's plumbing. That is well taken care of. She wanted to make a difference with issues like hot flashes, short term memory problems, depression, a sense of not belonging, loss of libido, sexual pain and difficulties, the lack of pleasure in everyday life. There came a point when she realized that she was not doing what she wanted to do and she began to express hatred at her job. You could see that she had given up. She had missed so many chances to help women that she believed that she had failed and that there was no way back. She rarely looked up from The Prescription Mill. You could see that it was easier that way. She was tired and the way back to her personal standards was just too hard.

Ethical principles are debatable. There is grey area between what I think my pharmacist's ethics require and what yours require. My ethics require me to give my attention to anyone who I think is in need. Having such stringent rules for my own behavior puts me in a difficult situation. How can I possibly help everyone who asks for my help? I have learned to discern the difference between people who just want my help and those who are, indeed, in need. The person who argues with me about the advice I have just given him is not in need. I know the difference. I usually ask that person why they are even bothering to ask me if they already know the answer. I also will not hesitate to tell them that the advice that I have given them comes from a font of education learned in school, CE and from many years of experience. If they do not want to take my advice, I have other things I need to do.

Most young pharmacists have gone without money for at least six years. The money they earn in retail can sustain them for a long time. It is easy to look away from ethical principles, put their nose to The Prescription Mill and collect the big paychecks. But not without a personal reprimand. We can't escape from ourselves. We are always there and we know when we have failed ourselves. There will be reminders and none of us can look the other way forever.

There is no debate about professional responsibilities. The argument, I can't counsel because of The Prescription Mill timers is not acceptable. To allow an Intern to conclude that they can break the law with impunity is unacceptable. Interns are too often mentored by pharmacists who have no business acting as Preceptors. Interns and young pharmacists watch veteran pharmacists flaunt the law every single day. None of us would ever consider allowing the pharmacy to be open when there is no pharmacist on duty. That is an administrative issue. That is breaking the law. Patient care is not involved. Neglecting to counsel is neglecting to provide patient care that is mandated by law. The young pharmacist's responsibility to counsel appropriately is a legal liability and they very well may be held accountable if they consistently break the law. There will be some highly publicized citations issued. Most likely in Washington State, Florida or Texas. What happens then will be a game-changer.

The three pillars should be enough, but there is much more that diminishes the young pharmacist's life force. The Interns are not pharmacists yet. They can leave the store and go to the local sandwich shop for lunch. They are free to enjoy the holiday breaks at their family's home two hundred miles from the school. They go to the bathroom when they have to go. The idea of a fourteen hour shift with no breaks is not part of their paradigms. Then, they become a Registered Pharmacist, take a job, and on the second Tuesday find themselves still at work after twelve hours with too much caffeine in their body and an empty pound bag of Butterfingers on the counter.
The cashier is gone for the day and the Technician has a dozen prescriptions to type. The young pharmacist tries to answer the telephone according to company policy and finds out that answering the telephone is all she has done for fifteen minutes. Never again, she promises, but she decides to forego counseling two hours of patients because of The Prescription Mill Timers. She confronts the Pharmacy Manager the next day. She complains that she is not practicing pharmacy the way she wants to. He shames himself by laughing at her and asking, "Do you think that Cost-Less Drugs cares what you want, Beverly?" The young pharmacist who does not document every single

incident like this is very unwise. They do not understand that the law is the law. They cannot ignore it and still maintain their integrity. The real questions is fundamental, one of context: Does Beverly work for Cost-Less Drugs? Or is Cost-Less Drugs the place where she has brought her practice of pharmacy for a price?

Dignity is something that you give to yourself. The Prisoners in The Prison have to fight to be dignified. The Jailers are relatively intelligent people. They would not be store managers if they weren't capable. They are smart enough to know how to get a young pharmacist's goat. They can spot a young pharmacist who is not complying as a Prisoner should and are likely to make getting the pharmacist in line a priority.

Some Jailers will play the company policy card at every turn. Some gleefully rub their hands together when they can play the "Rude" card. All of them will play the you-didn't-get-that-report-in-on-time card. They don't have to play the customer service card because that one is always on the table. The "write-up" is the club they use. The young pharmacist who can hold up against this withering attack by The Jailers is rare. They don't want to take a chance on losing their job, after all. Suddenly, they are in the same boat as the Prisoner with a family, mortgage, two car payments and kids in good schools. They can't afford to lose their job. They are Prisoners. The pleasure of working is gone. The difficulties they put up with are too huge. The "How" is too hard when the only "Why" is money.

If that isn't enough, you have Jailers who are very talented and creative and intrepid. They are like teenagers on dope. They come to believe that they are bullet-proof and they take chances that are not chances at all because the pharmacist is not going to complain to the people who can do something about it. Often, that is the government. The E.E.O.C. would be all over The Jailer who engages in sexual harassment, but is running his fingers through your hair actually sexual harassment? Is leering at your bottom and referring to you as "Lover Buns" when he is talking about you to the Assistant Jailer actually harassment? These are harassment and, if they put up with them, the young pharmacists are willingly giving away their dignity because they are afraid of The Jailers. They are given a club and they don't use it.

Young pharmacists do not enter the profession expecting the worst. The come in with rosy expectations of practicing their chosen profession the way they know it is supposed to be practiced. They soon realize that the divide between the job and the profession is huge. There is a slow erosion of their best expectations. They give up and just survive as Prisoners. Their expectations are shattered. What are they supposed to do when swimming out in the middle of the stream where the water runs clear is just too hard. They see everyone else wallowing in the muck. What do you expect them to do? Stand up to the Jailer all by themselves?

The veteran pharmacist who has given up and just can't stand the enthusiasm of youth should be censured. I watched a Pharmacy Manager tell a young pharmacist that he would be smart to just "Get with the program". He told the young man that his life would be better if he simply came to work every day expecting the job to be a "Piece of Shit". He suggested that "Making waves" would make the young man's life miserable. The young man did exactly as he was told. The excitement and purpose in his life was robbed from him and he became a Prisoner himself.

I have read complaints from hundreds of pharmacists. They blame the profession. I have heard from pharmacists who claim that they will do everything they can to keep their own children from choosing pharmacy as a course of study. It only takes a few years and some Prisoners start looking for a way to break out of Prison. There are teachers and real estate agents who are pharmacists. Drug Company representatives are pharmacists. There are stock brokers who are pharmacists. There are also plenty who left to build a career elsewhere and came back. The money is just too good.

Young pharmacists see all of these scenarios and, justifiably, ask themselves, "What did I get myself into?" They no longer think that they are different. They do not believe that they are too strong to succumb. I have heard from scores of young pharmacists who claim that they will make big changes just as soon as they have fulfilled the conditions of their sign-on agreement.

A number of years ago, one of The Governors (Executives) of one of the major drug store chains was the featured speaker at a meeting of some of the company's Pharmacy Managers. This Governor apparently wanted to show off and exhibit how well versed he was in Psychology 101. He was a Governor, after all, and a Governor's ego is often in the Masters of the Universe class. My friend in Seattle, the artist who wants nothing that smells of pharmacy in her life away from work, was in the room. This talk put this company behind by a decade in recruiting the best pharmacists. It is no wonder that my artist friend quit her job and emotionally bailed out. The Governor addressed Abraham Maslow's Hierarchy of Needs.

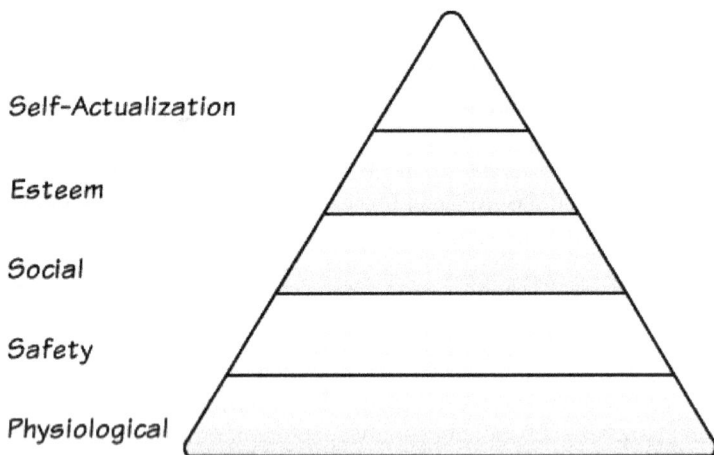

Self-Actualization

Esteem

Social

Safety

Physiological

## Maslow's Hierarchy

He started at the bottom and told the assembled Prisoners that their job made all of the basic biological and physiological needs at the bottom of the pyramid possible. The company paid enough money that the pharmacists could provide all of the food, drink, clothing, shelter, a place to sleep and warmth that they and their family needed. My friend reported that the executive made a leering crude comment about the pharmacists' sexual lives. Our artist friend said that he spoke in a loud, insistent manner as if he was daring someone to argue with him. He was a Governor, however, and even though there was a lot of uncomfortable shifting in the seats no one said a word.

The next level up is where the Safety Needs are listed. The executive insisted that the company paid enough money that the Prisoners could afford to live in a law and order community where there would be good police services to provide safety and security. He said that the pharmacists made enough to provide stability for their families.

He said that it wasn't the job of either pharmacy or the company to see to it that the pharmacists felt like they belonged to anything. He told them to join a church. My friend said that, at this point, The Wardens in the room seemed uncomfortable.

He spent an inordinate amount of time on the levels from Esteem Needs all the way to the top. He told these pharmacists that it was ridiculous for them to think that they could get self-esteem or self-respect from their jobs. That is not what a job is for. A job is to make money to provide for the bottom two levels.

This was the beginning of the end for our artist friend. She works part time now. She refers to it as chopping wood and carrying water. It is just about the money. She hates doing it, but the money is good for carrying water. She has beauty, balance and form in her life now. Her achievements are artistic and she has a reputation as an artist. There is personal growth and she does help others find their way.

How do you tell a twenty five year old that his professional life is not going to help him find knowledge and self-awareness? Too many pharmacy managers believe exactly what the Governor in our example believed. The pharmacist is just a worker. That is not what professionals believe about themselves. This is a toxic stew fed to young people who are hungry for meaning in their lives and want to know the way.

Thus, for the young pharmacist, begins the barren landscape of a career stripped of creativity, commitment and noble purpose.

There is not good enough "Why" to make up for a "How" when the Prison feeds unacceptable-to-any-other-profession behavior like, "Hey, you, where is da lawn chairs for nine ninety-nine?" or "You're just a pharmist. You can't tell me that these refills are no good just because it is six months". Just the proximity of the work area to the customers is demeaning. They put you right out front where anybody can get to you. Of course, the pharmacist is going to look at the young attorney seeing clients by appointment and charging in ten minute increments and think, What the…?

Dostoyevsky said: A man can get used to anything, but don't ask us how. It takes no time at all to become entrenched in Prison life. There are things that happen that can cause a pharmacist to almost lose all reason or they have none to lose. Thus, the illusions of an ideal life as a pharmacist are destroyed one by one. Most of us are overpowered by a grim sense of humor. It is necessary for our survival. There is nothing to lose anymore when we have so easily given away our self-respect and integrity for the money.

Have you ever listened to a group of pharmacists when they have had a few drinks at a social function? They fall into laughing about their situations. They play "You think that is bad, listen to this". Then they tell their story, trying their damndest to make it worse than the other guy's story.

They laugh as they tell about working all alone after a certain time in the evening. The clerk leaves at six and the last technician at seven and there are two more hours to go. It is raining and cars are lined up at the drive-through. The third in line starts honking and number two gets out of the car. He is a big man and he walks back to number three and stabs his finger at the closed window. Next thing you know, he starts pounding on the driver's side door of number three. There is no more honking after that. The telephone is non-stop ringing, but it has to be ignored. Answer the phone and you know that you are dead. They never give you a straight message. They always have a story they want to tell. Two lines are ringing now.

There is an irate woman at the counter. She tells you how long she has been waiting and you haven't even begun to work on her three prescriptions. A nice elderly man holds his hand up, "Take your time," he says. He hands you four prescriptions. You see that he has neatly printed his name, address, phone number and date of birth on each one. You want to kiss him. The floor manager makes an appearance. "Why are these people having to wait? They just want to buy their prescriptions." He huffs and puffs. He looks a little intimidated. He is a Jailer-in-Training. "Why are you just standing here? How long has the drive-through been waiting."

This is a modern pharmacy. White and tile, but it is now pure ugliness. The pharmacist experiences disgust and failure and hopelessness. The pharmacist's feelings, ambitions, standards, ethics and legal responsibilities have all been blunted in a matter of minutes. Walking out is not an option. The mortgage must be paid. The kids' schools. The club membership is new. The spouse makes a fraction of what the pharmacist earns. This means there is no freedom. The fight or flight instinct is smothered and the stomach hyper-acidity is worse than ever. The pharmacist gives up and survives until nine o'clock. Why is this funny? I almost expect them to break into song, like it was picking cotton.

The pharmacist is numb on the drive home. He is trapped. His throat is filled with the burning of the acid of disgust, horror and self-pity. He brushes past his spouse and ignores her greeting. He sees warmed up meat loaf on the table and barks at her, "I want a pizza. Order a damned pizza. How many times do you think I can eat meat loaf?" He takes a beer from the fridge and goes to his room. The game is still on. He kicks off his shoes, sits for awhile and then gets up for another beer. He is asleep when the pizza delivery arrives.

Two days later, he comes home at nine thirty and the house is dark. Inside, he finds that his wife and kids are gone. The closets are empty. Most of the furniture is gone. There isn't even a note. The pharmacist ends up falling to the floor. He is wracked with sobs. He rubs the mucus away from his nose. He is devastated at what he has caused. But, at the same time, there is a wave of relief that surrounds him. He can't do it with them, but he may be able to do this all alone. He can support them like he is supposed to. All he needs is a small apartment. He knows that he will be okay. Then, he realizes that he has to tell his mother and begins crying again.

That may be dramatic, but our pharmacist survives. The conditions of working in the pharmacy become so commonplace that after a year or two he is no longer moved. He has no problem ignoring the telephone when they take his help away and he is stuck working alone. He knows how to survive. If he chooses to keep the drive-through open, the drive-through customers wait just as long as the people at the counter. If they honk, they wait longer. He takes care of each patient in turn. The waiting time is two hours. He gives each prescription the same care he would give for an order for his ten year old. The floor Manager keeps away. This pharmacist has a reputation. The word is that he can be mean. When there were numerous complaints about his service, the District Manager came to the store. The store manager was with him. They stood aside, expecting the pharmacist to react, but he just kept working.

The Warden cleared his throat. "What seems to be the problem?" the Warden asked. "People complain every time you work the late shift that the wait time is excessive. You have to do better. You know the company policy on wait times."

The Jailer pipes in, "You don't answer the phone. You let it ring." The store manager is emboldened because the pharmacist district manager is doing the heavy lifting of reprimanding.

"Give me back my clerk and the technician and the wait times will be what you want."

"You know that we can't do that. There is a budget and we will intend to stick to it." The Warden stares at the pharmacist. "Everyone has to suck it up."

"I have no choice but to write you up." The Jailer is smiling.

The pharmacist has a new image. He wears a goatee. His hair is longish. He lives alone and his only stress relief is working out. He takes one step toward The Jailer and the puffy store manager retreats two steps. The pharmacist looks at the manager and smiles. "Fuck you and fuck your write-up, man."

The Warden stiffens. "You can't talk to him like that. I will write you up and you will sign it."

"Fuck you too, man." There is a pause. The pharmacist's eyes harden. "Fire me." He holds his hand up to silence The Jailer. "Do you know how many times you have told me that I counsel too much? How many times have you ordered me to speed things up by not counseling at all? I can tell you the exact words you used because I have documented every single incident ordering me to break the law."

The Jailer huffs, "I never... "

"Don't," The Warden stops The Jailer. "Don't say another word."

I know that this little fiction is extreme, but is it really that much over the top? It is a story that is at least partially true every single day. It is pathetic, but pharmacists don't have a say in how they practice their own profession. They just follow the rules. It does not matter that they are highly educated medical professionals. This is not right. They are Prisoners first. Dignity? What a laugh!

There are pharmacists who work an entire career like this. The conditions become so commonplace that the pharmacist is no longer moved by anything. The paycheck at the end of the week is all that matters. It is the only "Why" that counts. The pity is that The Prisoners are so driven by the rules that they have no problem throwing each other under the bus.

Another interesting sensation that occupies us is curiosity. Perhaps it is beneficial for our mental health. We actually have the ability to be observers of our situation. We are curious if we are going to come out of this okay. We are especially interested in how other pharmacists get along in their jobs and we want to know what is going to happen next to us. What is going to happen when the store manager finds out that the pharmacy is over budget in labor hours?

Will there be trouble when the inventory shows how overstocked we are? Will they give me my vacation in June? The family reunion comes only once every decade. It is my husband's family and he says that he'll take the kids and go to Iowa without me. How will that look? His aunts never did like me very much.

We find out that everything we held sacred is a lie. The state boards of pharmacy don't care that we are dangerous when we are hungry and tired. Pharmacy boards are mandated to protect the public from dangerous pharmacists and they don't recognize that every pharmacist in the twelfth hour of a fourteen hour day is a dangerous pharmacist. Can't they see what is happening? Pharmacists write to them every day and complain and ask the help of the boards. They answer that it is not their job to regulate working conditions, but can't they see that we are actually commenting on protecting the public?

Pharmacists make life or death choices all day long. After too many fourteen hour days in a row, as one company requires, the pharmacist is dangerous every single minute. Have you noticed how difficult it is to sleep after a long day without a rest period or a regular meal break? Some of us fill prescriptions all night long. My habit was to fill the same prescription, over and over, and never get it right. You go home at ten o'clock and go right back to open at eight o'clock. Is it any wonder that we make mistakes? It is a miracle that more people don't die.

In the end, we all are curious about each other and end up doing what we do every single day, many of us hating every minute of it. We diminish the seriousness of our dilemma. We laugh about it because if we didn't laugh, we would cry.

My last management job ended around 1996. A couple years before that, a middle-aged pharmacist who had been out of work stopped by the pharmacy to talk with me. He wanted to work part time. I told him that I had a regular part time pharmacist and that Mildred seemed to like the job and had not indicated that she wanted to leave. Her commute was an hour one way. She was always on time and she had never missed a shift. She was loyal and I would be loyal to her.

"Is she having trouble doing the job?" this man asked me. He gave me a knowing look that made me uncomfortable, like the look one might give you if he was trying to tell you about a wayward spouse.

"She is doing fine," I said. "Mildred is a good pharmacist." She was slow, but in 1996, the abject Prescription Mill craziness that we put up with now had not yet begun. There had been complaints, but not about her competence. The complaints were always about the wait times late in the day when she worked alone.

"Dave told me that you might be thinking of a change." That look again. What did he know? What was he trying to tell me.

"Dave told you?" Dave was store manager. "What did Dave say?"

"He just said that Mildred was having problems."

"Dave should mind his own business." This was coming from left field. I suddenly did not like this guy as much.

"You knew that Mildred has a drinking problem didn't you?" That look again. This time with eyes arched. This pharmacist was throwing Mildred under the bus, hoping to snag her two day a week job.

"She hasn't looked impaired to me,' I said.

"Have you ever smelled her breath?"

I kicked the guy out. I told him not to come back. The next day, Dave, The Jailer, jumped me. "Mildred is a drunk," he said. He demanded that I fire her and hire the hard core Prisoner who was not averse to using gossip for his advantage. He had tried to throw Mildred under the bus.

Of course, I refused, but I did have a talk with Mildred. I asked her if she had a drinking problem.

She released a long sigh and looked at me. "It never ends, Jim. That was in a previous life, during my first marriage. I've been sober for twelve years and four months. Who told you?"

I had no problem telling her who the man was. "He will never work here," I added, "Not while I am the pharmacy manager."

Mildred nodded. She thanked me. She stood tall and showed a self-respect and dignity that is often lacking in that environment.

Apathy is a way of life in the retail chain pharmacy world. Apathy is not caring and not caring that you don't care. The pharmacist doesn't think they can do anything about it anyway so it is better to not care. That makes it very easy to look away when our fellow pharmacists have problems. It also makes us very alone when we have our own problems.

There was an older man who had worked for a chain drug store company for thirty years. He was a competent pharmacist who had graduated from the typewriter and calculating price with a pencil and notepad era to the calculator to the computer. He was not comfortable with the computer and the computer in the 1980s was not user friendly. He took an inordinate amount of time to process new prescriptions. His problem was that he thought he could hurt something by not being perfect. The store manager fired him. He called him to the office and showed him pages and pages of complaints about his being slow. Customers said that he was grumpy and short with them. A doctor complained that this pharmacist refused to make a compound on a day when he had no pharmacist overlap. The store manager was up to something and everyone who worked in the pharmacy knew it. He had hired a new pharmacist before he even fired the older man.

The new hire was a pharmacist whom the manager had known for a few years. The Prisoner knew that he had given good service to the company for thirty years. He came to the two other pharmacists to enlist their help. He was desperate. The drug store was on Whidbey Island and the closest possible job was on the mainland, an hour away.

The other pharmacists shrugged and looked away when he talked to them. They changed the subject. They gave him vapid advice about where the jobs were. They were not fond of the store manager and they hated what he was doing, but they believed that they were helpless. They simply knew that there was nothing they could do to help our friend. They did not want to get on the bad side of the store manager, so they pretended as if everything was okay. They had no emotions about the firing. They were dulled and disinterested as if it couldn't happen to them. They knew that it was a sucker's game to care about anything anymore.

Pharmacists learn that they can endure just about anything as long as the paychecks keep coming. I knew a woman whose husband was mentally ill. He was schizophrenic and he was not much of a husband. This pharmacy manager and the divorced staff pharmacist began an affair one night when they were both tired and lonely. They kept this up for a year or two. The staff pharmacist was satisfied with the respite from his solitary life, but the pharmacy manager was consumed with guilt. Her husband knew about the affair, but that didn't help.

She was a woman who was raised in the south in the 1950s. When you married it was forever and you were always faithful. This woman's guilt was palpable on the days after the night before in a bed at an out of town hotel. She suffered greatly and demanded of herself that it never happen again. That was repentance and she was good until a few weeks later when it happened again. Her lover did not deserve such a loyal woman. He kissed and told and the pharmacy manager endured great emotional pain just from the knowing looks and the snide remarks. She endured though. Eventually, her lover became bored and moved on. It was for the best, she knew, but she still had a schizophrenic, house-bound husband and she cried at the loss of the diversion.

Would this scenario be different if she had been happy with her job? I don't know, but I'd guess that she would have had a better chance at managing her life in a manner that brought her some self-respect. She had not acted with integrity and she knew it. She castigated herself as no one else could.

The Prison has its own unique form of punishment. It is called the Write Up. The Jailer uses the write up when a customer complains about you. He will write you up when come in late to work. He will write you up if you do not follow company policy. I am certain that a company not named Three Pee Ex authorizes their Jailers to write up a pharmacist for not toeing the mark with the timers on the Prescription Mill. Conceivably, the store manager has the latitude to write up a pharmacist for just about anything he wants. That is a formidable club to put in the hands of a person who doesn't like you and may even hate you because of the money you make and your importance to the company.

I worked with a young woman who was a wonder. She was a Pharm D in the mid 1980s and she insisted that her nametag read Doctor. She worked her twelve hour shifts wearing a skirt and blouse and high heel shoes. She was energetic, beautiful and very smart. The Jailer hated her. He hated her because she didn't listen to him. He hated her because he wanted to date her and she laughed at him. He hated her because she warned him about sexual harassment when he ran his fingers through her hair one morning. He told her that there was something in her hair and he was just getting it out. The store manager wanted to ruin this woman.

One evening, two attractive young girls brought in a whole stack of prescriptions ten minutes before closing. The pharmacist was ending a twelve hour day. She was tired and none of the prescriptions needed to be started right away. She told the girls that they would be filled the first thing the next morning. The girls went ape shit. They used abusive language and the pharmacist refused to even talk to them. They were continuing the name calling and the use of profanities right up to when the pharmacist locked the door to the pharmacy.

The next morning, the pharmacist on duty thought it was odd when The Jailer asked for the prescriptions. He said that the girls had changed their minds and did not want them filled. The morning pharmacist did not say a word to the pharmacist from the night before. He did not want to get involved. A customer who had witnessed the altercation called the pharmacist at home and reported that the girls were bartenders at a strip club where the manager spent his evenings. It was a set up.

When the manager called the pharmacist to the office for his lecture and write up, the pharmacist accused him of setting her up. He denied it, of course, and then told the pharmacist if she wanted to keep her job, she would have to write letters of apology to the two bartenders. She told him to suck eggs, refused to sign the write up and went home early, thoroughly shaken. What was killing her was not the punishment for something she didn't do. It was not the write up itself. It was the unreasonableness of the entire situation, the injustice of it. To top it off, she had no allies. The other staff pharmacist told her that he did not want to get involved. The Pharmacy Manager refused to send an explanatory note to the district pharmacy manager. The Warden only knew that she had refused to fill the prescriptions. The circumstances were not included in the write up. It was the insult that was the most painful part.

There are rewards for being a good prisoner, one who behaves and follows all of the rules, even if they are wrong. The Jailer and the Warden might make sure that you get the vacation that you want while all of the other pharmacists have to fight for a week in the summer and piecemeal out the other three or four weeks throughout the year. The compliant pharmacist may get the best holidays off and never have to worry about having to work on his children's birthdays. I knew a pharmacist who played golf with the store manager regularly. They were real buddies and the resentment in the pharmacy was like a fog. When the pharmacist wasn't at work, the talk was vicious.

This pharmacist was almost like a trustee in the prison. The Jailer was all over the other two pharmacists about hurrying up, but he never said a word to the trustee and this guy was the slowest of the three. When the company held a summer picnic at a theme park, the trustee was scheduled to work. The manager changed the schedule two weeks before the event so the trustee could attend with his family. It didn't matter that the woman who all of a sudden had to work was a single mother with thirteen year old twins. She complained vociferously that it was not fair and the manager said that he didn't care. "Haven't you noticed?" he said, "Life is not fair." With that, he laughed and dismissed her with a wave of his hand.

There are punishments for being a bad Prisoner. Often, they are arbitrary and they represent a Jailer seeing what he can get away with. If the store manager does not like you and that is often the case, he'll try just about anything, based on a whim. The store manager that disliked me the most was a man in his fifties. He was an old hand and was used to being the lovable fatherly type to the female employees. He eventually was walked out of the store by security personnel for sexual harassment. A female pharmacist was the messenger who called the kettle black. She asked him one day how he expected to continue to get away with sleeping with both the bookkeeper and the cosmetician. Harassment was not what she was talking about. She sincerely wondered how he could be carrying on sexual relationships with two female employees at the same time. She believed that he felt threatened because she dared to voice what everyone wanted to know.

He began a campaign of harassing the pharmacist. One morning, he walked into the pharmacy and stood there silently, watching her work. She asked him what he wanted a few times, but, when he wouldn't answer, she just ignored him. The technician was clearly nervous. The pharmacist told her to just do her job and that to ignore the manager because he was harmless. I think that all she could see was a man having sex with two of her fellow employees and who could make her life miserable.

At one point, the Jailer said, "I see that you have a hard time keeping up. You have not had the counter clear for over thirty minutes. I'm giving you a half hour to get the counter clean or I am writing you up."

The technician took it personally and argued, "But they keep bringing in new prescriptions."

"I'm not writing you up, I'm going to write her up."

The pharmacist made the mistake of laughing. "You are writing me up for what?"

"For inferior customer service. You are making the customers wait unnecessarily." He was sputtering.

"Nobody has had to wait longer than fifteen minutes." She went back to work.

"That is too long. Company policy is no longer than ten."

She was strong. She was not intimidated, but those years when she worked with this store manager was when she started taking ranitidine. These kinds of mental conflicts are soul-destroying. The clashes of will-power that we have to tolerate are impossible for other professional people to believe.

This Jailer judged the pharmacists every single day, but his knowledge of pharmacy and what pharmacists do was limited to the rules of the company. He did not care that it was the pharmacist's job to make sure that a young mother was educated on how her four year old was supposed to use an albuterol MDI. He did not care that compounding takes time and that the average chain pharmacy does not have all of the supplies and equipment needed to do some compounds. He did not care that pharmacists have the right and responsibility to refuse to fill prescriptions when they are for #360 Norco-10, #180 Xanax 2 mg and #180 Somas 350 mg and the patient has driven sixty miles to try to find a pharmacist who will fill them.

"Well, the doctor wrote them. They are legal prescriptions. Fill them." He didn't want to hear about doubts that the drugs were not for a legitimate medical condition. To the credit of all of the pharmacists, they dug in their heels and refused to be intimidated. This store manager's ignorance about pharmacy was astounding and he was still The Jailer. To be judged by someone lower who has no idea of what a pharmacist's responsibilities are is upsetting to some pharmacists and depressing to others. I know pharmacists who take their dose of SSRI every single morning simply because they feel that they do not have control of their own lives. They feel helpless to change their circumstances. The future looks hopeless to them. They know that a professional does not wet their underpants, but it has happened to them and the Jailers complain loudly when the Prescription Mill comes to a halt while they are in the bathroom.

Every single pharmacist knows that lawyers take a lunch whenever they want. Attorneys charge for everything. Doctors are treated with consummate respect and deference and the pharmacist is referred to as "Hey You" and the questions are often in the category of "Where are the flip-flops on sale for ninety nine cents?"

Pharmacists are treated badly by The Jailers, the drug store customers, legitimate patients and each other. The Wardens know what is going on in The Prison, but they only provide lip service and, with their silence, condone the poor working conditions and the lack of respect for pharmacists. We all know that what we put up with is appalling. If you are unlucky enough to work for a non-traditional drug store company, it could be downright toxic. The Jailers in the big box stores and grocery stores know nothing about the pharmacy business and they could not care less about what you put up with. For non-drug store companies with only a limited number of stores, it is even worse. There are no strong pharmacist Wardens. The pharmacist is a disposable employee. They don't need you to make a profit. Some of them don't even want the pharmacy department in the store. The inventory is too high and your wage downright angers the Jailers. They diminish what you do and, frankly, treat you like an interloper, an unwanted indentured servant, if you will.

To make it worse, the high minded professional tasks in the Medication Therapy Management category are not available professional jobs for most pharmacists. When you do administer vaccines or manage diabetic patients, they do not make sure that you have the time and all of the important tools. They certainly do not shut down the Prescription Mill while you are with a patient in the counseling area. They don't pay you for it either.

I have never personally been so downtrodden and had so little confidence in the job of working as a pharmacist that I wanted out. But, I am an optimist. Of course, I have felt hopeless at times and I have been stupid enough to think that I can change things. Most chain pharmacists feel the same emotion at the end of the day. Relief that the day is over. They go home and, if they are polite, they tell the spouse to just leave them alone for an hour or so. Some are so angry and beaten down that they abuse their spouse.

We are often treated like just another employee by The Jailers, but pharmacists are the only employees in the entire store who are expected to work with zero errors. Prescription errors could conceivable cause great harm to the patient, but the store managers are more likely to come down hard on a pharmacist who makes ordering errors and is overstocked. The error of being over the labor budget is an even more egregious error. They practically do not even pay attention when there is a misfill, but can get irate when the pharmacy manager inadvertently schedules a technician for a holiday that ends up in an overtime situation. The priorities and the values are totally misapplied. And the pharmacist bites and suffers more when a technician gets a few hours of overtime than they do when they let a prescription for lisinopril get filled for lansoprazole.

I worked with a pharmacy manager who dispensed a non-drowsy antihistamine instead of nifedipine. This was in the 1980s and prescription volumes were low enough that the pharmacist often worked without a technician. The error was all his. A few weeks later, the patient, short of breath and pallid, asked me if the drug was correct. I told him the truth. The store manager did not seem to care that much about the error, but he was zealous about the write up. The pharmacy manager was more concerned about the black mark that would go in his record than he was about possibly causing damage to the patient.

I can't imagine any other profession that can cause as much cortisol to stream through the veins than pharmacy. Zero defects is an unreasonable expectation. However, we deal in a commodity and the evidence of an error is apparent. When a doctor makes an error in diagnosis, nobody has to know. When an attorney interprets the law wrong, it is often ambiguous and open for debate. Not us. If you dispense Seldane and the prescription was for Procardia, there is no ambiguity. There is no grey area. You are wrong. The Prison will punish you and, pathetically, they will punish you not because of the error, but because it is a chance to get back at you for some misty wrong that you have committed. Often, it can be because you make more money that The Jailer. Now, that is a crime.

There is so much riding on what we do or do not do and the professional choices we make that the citizens of the states have every right to expect that the state boards of pharmacy would make sure that pharmacists work under conditions that will allow them to be both physically rested and mentally sharp at all times. That is a joke that only pharmacists can really appreciate. It is no wonder that so many mid-career pharmacists still dream of getting out. They are Prisoners of Comfort, however, and most would not or cannot take a cut in pay.

When I was a very young pharmacist, well before there were computers and such a thing as a Prescription Mill with timers and reports on wait times, I hated it that I had to go to work five days a week and that I couldn't just up and go anytime I wanted. It did not take me much time to get over it because I knew that I liked the living I was making. I lasted eleven years before I took some time off. Lots of time off. From November, 1976 until May, 1981 I did not work more than twenty hours in any one week.

There was a stretch of almost eight months when I did not work at all. I was so tired of doing what the establishments expected me to do. Pharmacist friends hated what I was doing because they were too trapped by comfort and could not even imagine quitting their job and taking off. My mother and father and brother didn't like how I was living, but it was not their life.

My car was paid off, however. I did not have a mortgage. I had a year's wage in the bank account. My second wife was a young woman ten years my junior who had lived a hippie life and thought that what we had was pretty neat. If you want to escape the hum drum pharmacist's life, there is a trade-off. You don't have to work as a full time pharmacist to have your dreams come true. They just have to be a different brand of dreams. There are not that many jobs that pay what you are making.

You learn in grade school about the importance of being properly nourished and eating good, healthy food. You are a health professional and you counsel people on the value of a balanced diet, not too rich in useless calories or excessive amount of fat and sodium. We could hold debates among pharmacists about the demeaning message that a diet of Butterfingers, salty snacks, coffee and Diet Cokes tells you about what a pathetic life you have on the job. It is pitiable and we are wretched actors if we believe we have to put up with it. The store manager eats a nice lunch, unhurried and relaxed. You can count on that.

The most ghastly moment of some pharmacist's day is the first. Unlocking the pharmacy door and hearing the telephone already ringing causes some pharmacists' hearts to sink. When they see that all three lines are lit up, they are defeated before they even start. The technician is not scheduled until nine thirty and the pharmacist is expected to navigate the first half hour all alone. This is professional undernourishment. It is soul-destroying. This is when a pharmacist starts taking real estate broker classes in his spare time. Not because they have always wanted to show houses, but because they feel desperate.

A young woman pharmacist knew when she was in the parking lot if she was defeated. If the store manager's car was not parked, then the day looked okay. She started hyper-ventilating before she got in the store if the store manager was on duty. He didn't do anything but leer at her as she walked into the store. He was always waiting for her in the lobby of the store and he always undressed her with his eyes. He never did anything overt, but his covert harassment caused her to quit a job that she had ten years invested in. The pharmacy district manager tried to talk her out of quitting, but she would not back down. She was too ashamed to tell him the real reason. For some reason, she thought that the harassment was her fault.

What was most troubling was the attitude of the people this pharmacist worked with. They acted like survivors. They were relieved that they were not the ones that had to leave a good job. The pity was that they were so thoroughly institutionalized that they thought that what had happened was acceptable and normal. There was a complete lack of sentiment or empathy. If they cared at all about the pharmacist, they did not show it. This manager had every single employee isolated. They were not capable of standing together for anything. This is the condition that pharmacists often find themselves. On an island, all by themselves with nowhere to turn for support.

Then, a few months later, a new employee was very well aware that what was going on in the store was not normal. She made some phone calls and the pharmacist was invited to one of the other female employees home for a meeting. In attendance was a woman from the personnel department and two members of the Loss Prevention office. Two days later, the store manager was escorted from the store. It was too late for our pharmacist. The store manager had been with the company for many years. He got out with retirement. The company paid him off with a handsome severance package.

I have noticed something that interests me. No matter how bad the pharmacist thinks it is, they almost always get through it okay. Perhaps it is because of the money they make. I can't remember any pharmacist who fell through the cracks completely. Money is the great equalizer, I Guess. Money buys diversions and the smart rat whose job is a piece of shit, in their view, can always afford to go find fulfillment somewhere else.

I remember a pharmacist who traveled from bridge tournament to bridge tournament. He was a cerebral tiny man who did not have the physical tools to play men's games like golf or basketball. He did not like chess, although he was very good at it. He said that chess players were, for the most part, surly and unfriendly while the people who played bridge were easy to get along with. He lived in a small town where they played pinochle so he had to travel and he loved that aspect of the game. He even met a woman who became his girlfriend. It was a miracle because he was five feet two and she was a good four inches shorter than he was. One evening, I was teasing them about having to sit at pillows at the bridge tables. He laughed, but she just smiled and pointed her finger at me. "Don't tell anyone," she said, "I do sit on a pillow." What twinkling eyes. "It's a secret." This guy was perfectly happy with his job. He lived to play bridge. Like our artist in Seattle, pharmacy was simply something he had to do. It was just chopping wood and carrying water. This man had not found a way to simply survive. He thrived. Not all of us are that lucky.

I know a woman in the east who hated her job so much that she wanted to quit and go to work mucking out cages at a zoo. She loved animals, she claimed, and she believed that she would be fulfilled working for a very small wage doing something she thought she would love doing. I asked her how she thought she would like it if it was February and the lion poop was frozen to the cement and it was her job to scrape it up with a chopper and load it into the wheelbarrow and wheel out to the community animal poop pile behind the elephant house. She said that she hadn't thought about it that way and admitted that her fantasy only included spring weather. I asked her if she hated her job so much that she wouldn't be able to tolerate doing it part time and reminded her that a part time pharmacist job would pay more than double an entry level zoo workers job. That was all it took. I don't know how it turned out in the long run, but for a few years she acknowledged my contribution to her happiness. She did go part time. She did make enough money. Her second income was almost as big as her husband's first income. She was a happy camper, working three days a week and still getting benefits and living her life. Her promise to volunteer at the zoo fell by the wayside.

I know that the job can be lemons, but there are plenty of women who have made years worth of lemonade working part time while they had school age children. It is perfectly okay with them that they spend fourteen hours straight at the Prescription Mill. Probably because they only work two days a week, never on weekends or holidays and never on nights when their child is in the school play. I know many male pharmacists who hate their jobs even more because of this. They believe that the part time female pharmacist who is a mother gets preferential treatment. They are right, but, in the era of pharmacist shortages, what do they want? No part time pharmacist? These women are very smart rats and I, for one, say, "Go for it!"

I have heard stories of pharmacists bolting and finding a different way of making a living, but I can't think of one of whom I actually know who has pulled it off. How would you like to be a real estate broker in 2011? A man in Las Vegas sent me an e-mail a year ago. He asked me if a certain big box store was really as bad to work for as a pharmacist as he had heard. I had no idea, but wanted to know more. He had made a lot of money in real estate before the sub-prime fiascoes and after not making one penny in real estate for an entire year his wife got sick of him brooding in front of the television set. She gave him an ultimatum. Go back to pharmacy or get out of the house for good. He went back to pharmacy. He wrote me again and told me that he couldn't remember why he left in the first place. The filter before his eyes had changed dramatically. Come to find out, upon inquiry, the evidence is that this particular big box store is a terrific employer for a pharmacist. He lucked out.

For at least a few of the years when I was pharmacist in a Prison environment, I was virtually absent from much of my life. My marriage had been basically doomed from the start. My wife and I were too dense with needs to notice it.

To make it worse and allow some purulent festering, my situation at work dictated my relative emotional health or exacerbated the lack of such. I would come home from work and tell my spouse that I needed to "unwind" and asked her to leave me alone. I resented it when she told me about problems in her life. Didn't I have enough of my own? I dearly loved my small step-daughter, but I wanted her Mom to handle every aspect of her life. The child's father paid a measly seventy-five dollars a month child support when he paid at all. It was some kind of hippie arrangement made when money was not supposed to matter. I paid for my step-daughter's Montessori School and everything else. That should be enough, I believed. I did not care how much money my wife spent. My life was all about working and escaping from work.

In that marriage, I was absent sexually. I did it, of course, often reluctantly. It was almost like I had a grudge. I would do it with forced vigor and enthusiasm, but it was almost like masturbating with a partner. I just wanted the release and I wanted it hard with squeals and rapid breathing. I wanted it different every time and if there was any hesitation to experimenting, I'd get moody and pissed off. It was two people who had done it so many times that it was pure habit. We merely remembered how it was supposed to be and didn't really experience it at all. We knew how it would begin. We knew how we would do it, how long it should last and how it would end.

After, I would light a cigarette and she would dramatize how disgusting the smoke was and that she still could not believe that she had married a smoker. I'd lay there with a sweaty chest, drawing in huge gulps of smoke and exhaling into the air. I'd say things like, "Divorce me then" and "Kiss my ass." I wanted so much to tell her that I married her to be a father to her daughter, a child I adored, not because I loved her. It was pathetic and disrespectful. Emotionally, I was so consumed with being a Prisoner that I had nothing left for this wife.

I don't know if I can blame my job as a pharmacist for this dreadfulness. I know that it contributed to the misery. I am an idealist and my job was so far from how I thought how I should be treated as a professional. I was disgusted that I valued the money so highly that I allowed it to run my life. By then, I had a mortgage and car payments and two children, however. I had a new car. I enjoyed frequent trips to Bodega Bay, on the Sonoma Coast, or Lake Tahoe with a pocketful of cash for the Blackjack tables. It was a very shallow period of my life, for a very shallow man.

There was a time when my marriage was in limbo and we were separated due to my wife's straying. I took a mistress. You will be interested to know that she was a pharmacy clerk. There were no technicians in the day. She was beautiful and willing and, with her, my job did not interfere with a robust and healthy sexual relationship. She got everything I did not give to my wife. We were like two children finding a new game to play. There were rules and respect. The experimentation that I wanted was welcomed by this woman. This sexual relationship was a refuge and safe haven from the spiritual ghetto that was my authentic life. It didn't last very long she rolled over in bed one afternoon and asked me for a raise. She was a pigette and I was a pig. I fired her the next week.

I firmly believed that since I was tolerating such abuse and disrespect at work that I should get a free pass into the world of the doctor and his nurse, lawyer and his secretary. My life force was depleted by this affair. It was much too intense for a man with some normal sensitivities. I was guilty of being an adulterer even though I was not living with my wife. I was a man who was raised in the 1950s. Had I not engaged in this clandestine relationship, I may not have gone back with my wife and stayed with her for another twenty years.

There are a handful of male pharmacists on the west coast who were my fellows in this club. A pharmacy manager for a major chain confessed that he had a big problem. His job was difficult. He said that he got no respect and hated every minute at work. He had a wife and two teenage daughters. He was a Bishop in his church and that is another story because men who are Bishops do not do what he did. He supposed that he should be having a good life. He blamed pharmacy for all of his problems and said that he should have been a teacher like his mother and father.

He also had a mistress and a third daughter with her. He said that he loved his mistress and the new daughter was the joy of his life, but that he couldn't leave his wife. I asked him why not and he said that it would look too bad and that his two teenagers would hate him forever. In the end, this man, who could not choose, had to live life with no wife and no children. His mistress married another man and he was excommunicated from the church.

I believe that all of us men who looked elsewhere have thinking, feeling, suffering scheming minds and we were looking for that steam control. A willing woman was that release. I suppose that there are female pharmacists who have traveled this low road path, but I can't speak for them. I am not a woman and, unlike a few men, only one woman has confided in me. She actually threw herself to the shrews. She left her husband and two children for the band director at the local high school. She actually did an about face about her job as a pharmacist. It became her refuge. She ended up paying both child support and alimony. What other pharmacists perceived as horrors were tolerated very easily by this woman. What she feared most was loneliness and not being able to make a living so she could still be a good mother every month when she wrote out the checks. She expressed that she was surprised how giving money to her ex-husband soothed her guilt. She did not think that was possible. She finally understood how so many men felt.

I really liked this woman and spent some time talking with her when she needed to talk. We did it on the telephone because my second wife expressed jealousy. This woman told me everything. The affair began when she missed most of a band concert because she had to work. Her daughter iced her out and refused to talk with her about it. She confided in the band director. He invited her to have a drink with him and, since her husband was out of town, she did something she would never do if she was not so emotionally beaten down. She accepted.

She was a Prisoner at work and a Prisoner at home. She put up with abuse at work and abuse at home. Her husband loved the money she made but complained about her job. He was a bully and the fact that she made more money than he did just exacerbated his meanness. She couldn't do anything right. If she prepared meals in advance, they were never what the family wanted to eat. If she did not cook, she was accused of inattention and neglect. If she did not clean the house, the husband complained. When she hired a cleaning woman, the husband said that she was wasteful. She couldn't win at home and she couldn't win at work. The pity was that she believed that this was a normal marriage relationship. One sided. The pressure was too much and what eventually broke down was not her job. She kept working, but what fell apart were her wedding vows.

The band director was not especially handsome or charming, but he was attentive and supportive. He just listened, she said. At the third furtive meeting at the cocktail lounge of a hotel in the next town, the new couple graduated to a guest room upstairs. She had never been with any man other than her husband up until that night. She told me that she was at first uncomfortable because the band director was so gentle and so patient. She was used to a quick, grunting business and never expected to get much pleasure from it. That changed that night.

I will never excuse cheating, but I can never justify a husband who is a boorish brute either. So, I'm not going to be a judge in this matter. I will just say that tolerating a brute, be he a husband or a store manager is never a requirement of a job or a marriage. Never mind about the husband. I can't vote about this woman's husband because I have never been a wife. I love this woman like I would love a sister, but I don't doubt that she embellished her stories about her husband, but I don't doubt for one second the substantive facts were true.

I have been a pharmacist for a long time, however, and I reserve my right to vote about a store manager who mistreats a female pharmacist. A man with a little bit of power can be like a hyena smelling blood when he finds that a female pharmacist who makes more money than he does has some weaknesses. Some managers will act like a Jailer with any pharmacist. Can you imagine with what kind of zealotry he can go after a woman who shows any kind of fragility?

Imagine how a real male Jailer would act when his real prisoners are women. Now imagine that the Jailer is a drug store manager and not that satisfied with his own situation and needs a scapegoat. Imagine that his own wife is not that happy with him. Imagine that his children give him back talk and that his wife defends the kids. Imagine that his mortgage is too hefty and that his wife quit her job as a clerk at an insurance agency and she still expects him to pay for the children's extracurricular activities. Imagine that she turns her back on him and calls him a loser when he rolls too close and tries to kiss her on the neck. Imagine what he sees when he looks at the female pharmacist who has never stood up for herself. He sees red meat to be pounced on.

Imagine that all of what I have described is true. Imagine that the poor bastard has been emasculated. That's too damn bad. I say pull the plug and let him go sell cars. A drug store manager who mistreats any pharmacist gets my vote to be put on the no call list when the bonus checks are handed out. However, I love women and I was trained by my mother to do my best to protect women in need. For any manager who mistreats his women pharmacists, I have one vote and that is off the island for good. A manager who mistreats any vulnerable woman is the quintessential pig.

That does not diminish for one second what pharmacists have to put up with in a Prison situation. We work strange hours, often until late at night. We come home enervated, hungry for both food and some company. Often, we find neither and have to fend for ourselves. We live in a cultural hibernation. For some of us, the only real social life we have is through our children. We go to their games when we are off. The school plays and concerts are important so we make an effort to arrange to be off. Other than that, we don't do much.

The best advice for any woman pharmacist, or any male for that matter, is to always look your best. Appearances are everything. If you look worn out and too tired to do your job, it will be easier for the store manager to mistreat you. Wear fresh clothes and make sure that your white jacket is laundered frequently. Don't give the Jailer anything to use against you. He will have no trouble finding ammunition. You don't have to help him. Get your hair cut when it is too long. Follow company policy on facial hair and piercings. Wear adult shoes. Sneakers are okay, but the white ones with springs on the heels look like children's shoes. Walking shoes are good, all one color.

I was hospitalized with polio in 1951. Because of that illness, for decades I have limped when I get tired. In the mid 1970s, before sneakers were socially acceptable for the pharmacy environment, I wore hard soled oxfords. My limp at the end of most days was noticeable. The night before I went and bought a pair of black sneakers, the store assistant manager stopped me at the door. This man was destined to become one of the most rigid store managers I ever worked with. He stopped me and said, "Jim, you are really limping badly tonight. Are you sure you can do this job?"

I was taken aback and actually stuttered trying to say, "Of course I can do the job." Immediately, I became concerned. Could they take the job away from me just because I limp when I am tired? Would my district manager stick by me? Would the store manager make my life miserable when the assistant told him about this? I worried unnecessarily about this for days. I was institutionalized. I was a Prisoner in a Prison environment and the two managers were the Jailers. It was totally unreasonable for this incident to ruin my peace of mind, but I could not see that. A prisoner in a real prison, with real jailers would tell you that they know exactly what I went through.

In the early 1980s my family belonged to a California swimming pool association. We joined because my step-daughter wanted to be on a swim team. We used the pool for swimming a few times a week and spent time with books in the sun or in the shade under the trees, but the swim team was the anchor. It is pathetic, but the swim club, swim meets and work were my life. I rarely went to the movies and I love movies. It was even more rare to invite my wife out on a date for dinner. We lived in one of the most dynamic spots on the planet, the San Francisco Bay Area, and we may as well have been living in a provincial backwater fifty miles from Idaho Falls. I was culturally deprived by my own lack of interest in life.

I love the ballet, the symphony and the theatre. The last show I saw was A Chorus Line in 1975, the New York cast in San Francisco and I had to be dragged to that. Before that, it was Hair in 1969. I was a Prisoner before it was fashionable for pharmacists to be victims. I believe that I was institutionalized before there was an institution. Perhaps, I am not alone. Could it be that pharmacists are emotionally suited to this type of life? If they were otherwise, would they have become doctors?

There are those among us who have a devil-may-care attitude about it all. They are the ones I looked down upon for years. They didn't seem to care about anything. They didn't care about all of the markers that made a chain pharmacy successful in the eyes of the Wardens.

Back in the day when one hundred dollars for a bottle of anything was a lot of money, a relief pharmacist bought a new antibiotic for over one hundred and twenty dollars for a bottle of 100 tablets. The prescription was for twelve tablets and the doctor was from out of town. The chances of using this medicine before the expiration date was between slim and none. I asked him why he ordered it.

"Because the customer has a prescription for it."

"But it will go out of date and we'll lose a lot of money. Why didn't you just tell him that you couldn't get it?"

"Because he needs it." He looked at me. "It's not your store, man. They sure have your soul, don't they? You are a pharmacist. Don't forget. You are a pharmacist." He watched me for awhile. I must have looked lost because his tone softened. "You take this too seriously. They are not going to fire you for practicing pharmacy and if they did you can find another job tomorrow."

"I like this store. I have been here a long time."

"You like it too much. You are habituated. It's like a drug. Maybe you should just quit and go out and find your soul again."

I don't think any of these pharmacists who were actually free from the institution were pharmacy managers. Some of them had been managers in the past, but had bailed out for one reason. They were not willing to put up with the artificial demands of the business. It certainly was not because of professional reasons, because they were professionals in every sense of the word. It was simple. The job depressed some of them. They stepped down and became staff pharmacists.

One man was single and said that he got so sick of the company run around that he found his dream job after he quit the manager position and took a job with a grocery chain. He was a floater in a rocky mountain state. The towns were far apart so the jobs were far apart. There were days when he would work until nine at night and be expected to show up at nine the next morning at a store two hundred miles away. His shifts were always twelve hours and he was paid time and a half over eight hours. The company paid him for his hours on the road and paid all of his expenses for five days a week. Meals, hotel, laundry and he had negotiated an additional one hundred dollars a day per diem. When he talked about it, his eyes gleamed. He loved his job and I could see why when he explained the reward for all of this work. It was Labor Day weekend and he told me that he would work only until the first week in November and then he would take five months off and head for a place on the beach in Costa Rica.

"How can you afford that?" I asked him. "Five months off?"

He smiled. "I get tired, but I don't even have an apartment. The company pays for my hotel room every single day I work and that is seven days some weeks. I pay for the others. It's the end of August. My year started the first week in April and I've already made over one hundred and twenty thousand dollars. That's in five months. I have no family." He looked at me and saw something. "Yes, I work a lot, but those twenty weeks off every winter on the beach are golden." His eyes widened. "I mean on the beach. The place I rent is on the beach under some palm trees. The closest neighbor is a hundred yards away. The village is a half mile down a dirt lane. I spend every evening in a little bodega with the locals. We drink Mexican beer. I even have a girlfriend down there." He shrugged. "If she gets bored with me or gets married, there are other girls." He told me that he would be at this one store for at least six weeks. The pharmacy manager had fallen off her horse and broke her leg. "Compared to my days as a manager, life is good." He winked. "Very good."

To a person, every pharmacist who was happy with their job as a pharmacist had, at one time, been very unhappy with their job. They seem to be able to do the job their own way and to not give any consideration to the expectations and demands of the institution. Most of the happy pharmacists have gotten to the point where they don't care if they are reprimanded for not making the numbers. They practiced pharmacy at their own rate in a safe manner.

One man in Vermont takes a half hour during a nine hour shift to have an uninterrupted meal break. He leaves the pharmacy and takes his brown bag to the office. He puts in the ear pieces of his iPod and listens to music as he eats a leisurely lunch. He reads the paper and always calls his wife to see how her day is going. No one has ever criticized this behavior after he made it clear that the only way he would quit taking a half hour lunch is if they fired him.

I work with a pharmacist who routinely closes the drive-through when the evening rush comes. So far, no one has complained about it. A woman in Washington State unplugs the telephones because she knows that if she answers the phone she is liable to get waylaid by a patient who just wants to tell a story or ask questions incessantly. She figures that a patient pissed off about the phone not getting answered is better than a livid patient who will complain about her being rude because she has to blow them off.

There is a pharmacist in California who got sick of the restrictions and rules of retail and now makes his living as a psychiatric pharmacist. He has the authority to write prescriptions after the psychiatrist or psychologist does the diagnosing. He put in many hours to become certified to do this job and he loves every minute of it.

These people understood that the conditions that were imposed on them in retail were not normal for a professional. The conditions are, however, very normal for the institution of retail pharmacy. All of the things we hate so much are normal. The ridiculously long shifts that are required at some stores. The lack of rest periods, for meals or just rest. The constant supply of beef jerky, sugary candy or salty snack bags on the counter with a steady diet of caffeinated beverages. The standardized equipment that is an ergonomic nightmare for 80% of the pharmacists. The fight a pharmacist has to engage in just to get a vacation in the summer. Hours and hours on our feet when we know that our legs will eventually be worn out. Bullying store managers. Sexual harassment. Age discrimination and gender discrimination. A condition where a tired pharmacist is dangerous and where every day the health of patients is in peril. All of this is normal.

A psychologist once said, "There are things which must cause you to lose your reason or you have none to lose."

In many cases, our reason is absent after we become institutionalized. An abnormal reaction to an abnormal situation is normal behavior. All of those things that we hate so much are perfectly normal because we are institutionalized and because we tolerate them.

New pharmacists are often preoccupied with a sort of longing for the ideal of professionalism. Then they become disgusted with the reality of what it really is. You can't continue like that. There has to be a killing or mortification of normal reactions to the conditions. The pharmacist becomes numb or anesthetized in order to survive. In the beginning, after being institutionalized, the pharmacist will look the other way when a fellow pharmacist is treated unfairly. Later, he becomes a disinterested, helpless observer. The pharmacy manager is about to be written up again for being over his budget on technician hours. The staff pharmacist does not care, feels nothing about this. It is just the way it is. I worked with a pharmacist who had limited skills, but he worked very hard. He showed up on time and did his job. He was slow, however. When the store manager fired him, I was called to the office to witness the firing. I watched, without any emotional upset, the scene that the store manager had prepared. It was a hatchet job. The fired pharmacist left the office and the new pharmacist that the store manager had hired walked in. I felt nothing. Life went on. I didn't feel shame about that until years later when I happened to run into the guy who had been fired. He was working for an independent. He asked me why I hadn't stood up for him. I had no answer. On retrospect, my lack of emotion surprises me. I am not that kind of man. I was institutionalized, but is that a good enough excuse?

The necessary protective shell is made up mostly of apathy. I keep returning to this theme because it is central to the pharmacist's predicament. The blunting of emotions and the accompanying feeling of helplessness is what allows for abnormal situations to be accepted as normal.

When I was a manager, I was paid a salary. Our base hours were 40 hours a week. We were short staffed and had difficulty getting the warehouse order on the shelves in a timely manner. Even if my shift did not start until the afternoon, I was in the habit of coming to the store and marking and shelving the order as early as six in the morning. It was a foolish, childish loyalty to the job. Oh, what a good boy was I.

One morning, there was a knock on the pharmacy door and I let the store manager in. He asked me what I thought I was doing. I told him that the job had to be done. He laughed at me. He told me what an idiot I was for putting in extra hours for no pay.

What really bothered me was the injustice of his assessment. It was unfair of him to belittle me. It was the worst of insults. I thought that I was being a good manager and I became the butt of a joke. Other employees knew that I did this, but it had not been a comic source of laughs until the store manager made it a funny story at my expense. I was institutionalized.

It is interesting to me that I don't know very many pharmacists who are religious. I would expect that people who feel in a hopeless situation would look for guidance and comfort in organized religion. There be many, but I don't know them.

After working in these conditions for awhile, reality dims and the overriding efforts becomes centered on one task. Getting through the day is all that matters. It is typical to hear pharmacists who work the early shift to sigh with relief and say, "Well, another day is over." I don't think that I ever asked any pharmacist, "If it is that bad, why don't you quit?" I knew the answer and I didn't want anyone to ask me why I didn't quit. I was a slave of comfort and was fearful that I would have to give up too much if I quit or was fired for refusing to comply with the rules of the institution.

Fear is not a good reason to do or not do anything, but fear is a powerful motivator. If you fear that you will not be able to provide the basics for your children, you will toe the line. If you fear that you will lose that nice new Lexus, you'll tolerate the day to day indignities that all pharmacists put up with and are part of the job. If you fear that you will have to give up some of what you consider to be the nice things in life, like the season tickets, the membership to the club or the new boat you'll let your ideals bend. If your spouse depends on your salary to achieve a certain status in the community, you may fear losing that salary. You will notice that we are discussing a fear of loss.

I was a relatively well off pharmacist in the late 1960s and early 1970s. California was a brain-drain state and pharmacists made more there than in any other state. My wage at my first job in California was 270% larger than the wage at my job in Ohio. My first wife loved the money and she loved the status that it brought her. She had more money to spend as she wished than any of her friends. The one who was the wife of a Ph.D. Chemist for Dow Chemical was green with jealousy. I was young and I took some pleasure in this. After a few years, however, the marriage was not working out and I realized that if I could not continue making the big money, my wife would be unhappy. It followed in my thinking that if she was unhappy, she would go out and find someone else with money to have an affair with. My job became more about conforming with the program to make sure that it was secure than about practicing pharmacy. I stayed in a job I did not like because I feared what would happen if I didn't bring home the big paychecks. I was living in fear of losing my wife. She did have affairs, by the way. She had affairs with many rich men and it didn't matter how much money I made.

In the end, it is fear alone that keeps us Prisoners of Comfort.

*You gain strength, courage and confidence by every experience in which you really stop to look fear in the face. You are able to say to yourself, "I lived through this horror. I can take the next thing that comes along." You must do the thing you think you cannot do.*

*Eleanor Roosevelt*

The situation that thousands upon thousands of pharmacist find themselves in is not hopeless. If they see themselves as victims who are powerless in their own lives, they are most likely going to be miserable. However, if they see that they have choices at every turn along the way, there will always be hope.

I have done my best to illustrate the worst pit that many of you wallow helplessly in so that I can show you that you are not helpless and that the difference is a matter of perception, pride, dignity and integrity. Pharmacy will not make you proud. Your viewpoint of yourself and how you behave is what drives your self-respect or lack of such. It is not easy to intimidate a person with confidence and self-respect. Especially, a professional. The relative intelligence and education of the average Jailer cannot compare with yours. He knows that and does everything in the book to keep you controlled.

The sense of being powerful in our own lives is largely missing with the institutionalized pharmacist. Like a disaffected teenager, we compensate for a lack of power in our professional lives by seeking power in ways that are not necessarily healthy. For every female pharmacist who finds some power by volunteering at the hospital for children with neurological disorders, there is a pharmacist who is depressed about her life and stays at home on her days off, lying on the sofa, watching soap operas. She is so used to salty snacks that she can't stop. There is an empty large bag of chips on the floor and a half filled 2 liter bottle of soda. She has gained weight and hates herself for it.

For every male pharmacist who is on the Board of Directors of the local school board, there is a man who gets his power the way I did, from controlling my body to the point of distraction. I cross trained every single day. I spent an hour or more on the Nordic-Track machine six days a week and swam laps four times a week. I restricted calories to the point that I was so lean that I actually looked like I was ill. A friend of my brother expressed his condolences. He said, "I'm sorry about your brother." Mark said, "Why? What's the matter with my brother?" The friend said, "He's got AIDS, doesn't he?"

What I got was power over my own life. I may have been a victim at my job, but here I was in control. Nothing was more important than my physical condition and my lean appearance. I cringe when I see pictures taken back in the day because I actually believed that I looked good.

# Recovery is regaining Your Power

Essentially, the miserable pharmacist is wretched because they choose to be unhappy. There is a choice every single day to be proud of what they do or to blame the job because they are not happy. They don't even use the best tool available to them. That tool is anger!

Anger is fuel. It is not the bad thing that your parents said to suppress as mine did. "Jimmy, nobody needs to know you are angry. You should control yourself." We feel anger and we become frustrated when we hide it because we want to do something about it. This goes against the image of the calm, in-control professional. Instead of showing the anger, we stuff it and chug Maalox and take two 20mg omeprazole every day.

How would it look if we showed that we were angry? At work, you don't hit that someone or break that something or throw that fit. If you smash that fist against the wall, do it in the bathroom where no one can see that you are out of control.

What we do with our anger is deny it. We stuff it so far down that we forget what makes us angry. We are institutionalized and we believe that we should not get angry. We lie about being angry at the store manager. We hide our anger at the lack of technician help. We do not express our outrage to the district manager. Doesn't he know that it is his precious customer service that pays the price?

Some of us hide it so well that we medicate the anger and filch the occasional lorazepam to hide it even better. We are professionals and professionals are nice people. We bury our anger. We block it and we hide it.

What we do best with our anger is lie about it. Unfortunately for our spouses, we lie so well that we often take our misery out on the people we love (or are supposed to love) the most. We do everything but listen to our anger.

Listen to your anger. That is what it is meant for. Anger is not a polite request. Anger is a scream. It is a command. It is a slam of the fists down on the table demanding your attention. Anger has a right to be heard. Anger should be appreciated and valued. Anger must be listened to if you are to regain your professional balance and power. Why? Because anger is an atlas or a chart or a diagram back to living the ideals you had when you were in pharmacy school.

Anger reminds you of your boundaries and limits, the areas where no one was allowed to tread without your permission. If you can set up the periphery of your professionalism in just one area, more will follow. If you list only ten serious drugs that you will counsel on no matter what, your list will be twenty in little time. If you let the store manager know in writing that his touching you at anytime, in any manner, is unwanted, you will regain enormous power and control over your own life on the job. You can gain power simply by refusing to get wet underpants because you neglect going to the bathroom when you have to go. Documenting anything at work that makes you uncomfortable will give you surprising control.

Anger shows us where we want to go. We may not know exactly what we do want on the job, but our anger tells us, without ambiguity, what we sure as hell do not want. That is a really good place to start because anger shows us where we have been and sets us on the course of recovery. Anger is not a sign of disease. It is a sign of health. If you no longer get angry at being institutionalized, stop, take a deep breath, and examine how you will find your way back. I contend that you will find that the first sign of recovering your health, well-being and pride will be anger. Welcome it. Savor it.

It is not very healthy to act out from anger. That is childish and not productive. I quit a job once out of anger. It was a good job. I was well respected in the community. The problem was that the store manager tried to micro-manage my department. I have never bent to management from a non-pharmacist. This guy was out to bring me to my knees. I fell right into the trap. I became so angry that I brought the problem to a head with some stupid brinksmanship. My district manager did not back me as fully as I wanted, so I quit. My one-way commute for that job was less than ten minutes. The one-way commute for the next job was ninety minutes. I was like a teenager having a meltdown. I turned my anger into indignation without any examination of the circumstances. I was an idiot.

Anger is there to be acted upon. Anger points the direction. Anger is the wind for our sails as our sailing ship tacks as we move on the appropriate bearing where our anger guides us. Had I used my head and had the presence to translate what the anger was telling me, I would have made better choices.

"Damn it, I could run a better pharmacy than that!" This anger says that you want to have your own pharmacy, you just need to put all of the pieces together.

"I can't believe it. Mildred told me that she was going to demand a transfer to the suburbs and she got it. That's what I wanted." This anger says: Stop keeping your goals and dreams hidden. You need to express your wants and believe that you deserve your dreams to come true.

"That was my idea. This is unbelievable. I mentioned it only once and that son of a bitch took my plan and put it to work. He gets all of the credit and I get none." This anger says that it is time to take yourself seriously and show yourself some respect. Your ideas are good enough to do something about.

Anger is the tornado that blows away all of the restrictions and hesitations and lack of self confidence of our old lives. Anger is a valuable instrument to be used productively. Anger cannot be the master, only the servant. Anger is a deep well of power, if used properly.

Apathy, laziness, misery and gloom are the enemies. Anger is not a good buddy, but anger is a friend. Not a mild-mannered friend, but a very loyal and steadfast friend. Anger will always remind us when we have been cheated or cheated upon. It will always tell us when we have been deceived or when we have betrayed ourselves. Anger will tell us that it is time, finally, to act in our own best interests. Anger is not the action itself. It is the action's invitation.

# Watch out what you ask for

You might just get what you want and then what are you going to do? It can be scary, having dreams come true. That means that you have to take responsibility for your own life. This is not comfortable, but you will feel the power. You can no longer blame the big bad store manager wolf for your lack of integrity. You can't say that the company made you do it. You are back in your own hands, a professional making choices every day that benefit you and your patients. This is a good thing, don't you think?

When you take responsibility, things happen that you cannot fully understand why. You are the pharmacy manager and you tell the Jailer that you are going to do what is best for your department in all business and professional matters. You tell him that you are the pharmacist, that you know best and you request that he mind his own business. You take all appropriate actions, the department thrives in all areas. Your pharmacy is suddenly the most professional and most profitable in the company and everyone wants to know why. What did you do?

Taking responsibility is not easy. You can feel very much alone. It takes courage to do the right thing. This is a difficult and slippery slope. A pharmacist who has little self-respect and has been stripped of dignity may need assistance in making the choices that are best for both their professional and personal lives. I honestly do not think that you should rush. You have been institutionalized for years. There is no hurry. You don't want to make a rash move as I did. You probably should not try to do this alone. Talk to someone you trust before you take any significant action.

*I merely took the energy it takes to pout and wrote some blues*

*Duke Ellington*

# There is a way back

Deborah Van Sant is a pharmacist in Arizona. This is a snow bird state with lots of pharmacies competing for the baby boomer trade. Deborah Van Sant opened her own pharmacy in 2010. She told me, "If I only had one word to describe how I'm feeling about my choice to open my own store, it would be LIBERATED."

"I have a license to practice and I am now using that license as I have been trained to do. I no longer have to compromise my ethics. I no longer have to put profits above safety. I no longer have to work 14 hour shifts!"

Deborah gave a talk to more than one hundred students at the University of Arizona, College of Pharmacy. Most of the questions at the end were not about the subject. The students wanted to know all about her opening her own store. Deb's store was suddenly a sought-after rotation site.

Deb said, "The chains are our best advertisement."

Nathan Schlect owns a drug store in North Dakota. He does very well. North Dakota requires that pharmacies have majority owners that are pharmacists. There are no Rite-Aids, no Wal-Mart Pharmacies. No pharmacies in grocery stores. Enough said.

RxJoe owns a pharmacy in Boston that is so close to a Rite-Aid that you can read the sign from Joe's front door. There are two CVS stores within a mile. One where he was the pharmacy manager. He started his store with a compounding laboratory. He had to work hard in the beginning. Long hours. RxJoe is a good businessman as well as a good pharmacist. He told me that his compounding laboratory and unit dose business for nursing homes are the winners.

Rich and Marge McCoy are a man and wife pharmacist team. They own the Lopez Island Pharmacy in the San Juan Islands of Washington State. The store is a full-service drug store in a small village in one of the most beautiful and idyllic locations on the planet.

I happen to believe that opening your own pharmacy is not just an airy-fairy idea. I believe that, at this time, having your own pharmacy is a viable and profitable alternative to working for your living at a grocery store, big box store or one of the chains. Pharmacists, by tradition and by law, are the medical professionals who dispense prescriptions. We are always there. It's the law.

Think about this: In 1999, Americans spent 104 Billion dollars for prescription drugs. Prescription spending increased to 234 Billion dollars in 2008. That was an increase of 130 Billion dollars in just nine years. Moore's Law of exponential increases tells us that it will not be that long until pharmacists fill One Trillion dollars worth of prescription in one year. That is one gigantic pie. Why shouldn't you get some of it?

Among people older than 60, the researchers from the National Center for Health Statistics reported that 88% are using at least one prescription drug and 67% are taking five or more prescription drugs. That's today's figures. How many will take ten or twelve in Rx drugs in five years? 10,000 Americans turn 65 every day.

The baby boomer represent the largest and richest consumer group in history. The first boomer to get Social Security pension payments filed on October, 15, 2007. The baby boomers will not age gracefully as the generations before them did. They want to remain youthful, strong, active and sexy for the entire ride.

Big Pharma is committed to helping them get what they want. Viagra was the first drug designed for the boomers. It is for relatively healthy people. The bank is choking with money that the boomers will spend on their vitality. Watch for seligiline in low doses every other day. A doctor in The Czech Republic touted seligiline as an anti-aging potion twenty years ago. It is already all over the Internet. Pharma sees the opportunity for landmark profits serving the baby boomers. Big Pharma will be driving the train. In 2010, they began to promote testosterone replacement therapy to men by advertising a condition they called "Low-T" on television. Why fight it and tell me a good reason why you shouldn't get a slice of this pie?

It is conceivable that a pharmacy practice based on serving only the boomers will be able to provide a good living for the owner. However, no good businessman would exclude any group. The population of the United States is over 311 million and growing. Pharmacists may see that the need for prescription drugs may be driven artificially by the manufacturers, but are you going to go on a Don Quixote mission? Are you going to tell the mother of a member of the 6% of all American children who use drugs for ADHD that the condition is a myth and all they need to do is give their child the attention they need. Are you going to let someone else fill those prescriptions?

I don't think that an independent pharmacy will fill the bank filling prescriptions. Not with the ridiculous contracts the PBMs demand these days. There is a lot more to be done.

Let's focus only on the baby boomers. They will have needs that will be begging to be filled. A niche or "boutique" business that services the needs of aging Americans will thrive. Nutritional support is particularly important to an age group that cannot get all the nutrients needed for good health from their diet.

For starters, the baby boomer women have needs that are not being met by anyone. These people get perfunctory attention from their doctors. With a little bit of education and planning, a female pharmacist could become a Women's Health Counselor. A woman's plumbing is not what needs attention. That is well attended to. What the baby boomer woman wants is help with hot flashes, loss of short term memory, lack of libido, loss of vaginal integrity, lack of interest in life, sleep disturbances and depression. Menopause is difficult. The woman is no longer young and she knows it.

Compounding can be a niche business. I am acquainted with a pharmacist in Vermont who does all sorts of compounding in his laboratory. He specializes in bio-equivalent hormone replacement therapy. He sends saliva samples to a laboratory in Oregon, analyzes the results and consults with the patient's physician to determine the appropriate formula. Giant Eagle Grocery Stores recently purchased an independent and will make it into their compounding pharmacy. Why does that bother me? A grocery store? Walgreens has compounding centers. Why not you?

Making your store a center for immunizations can give you a solid clientele and enhance your professionalism, reputation and standing as a member of the health care community. I stopped into a small, vintage drug store in 1999. It is in the Fairhaven District in Bellingham, Washington. At first look, it was a crowded, old-fashioned store, very old-fashioned. Then, I noticed that he advertised every immunization there is available. He was a front line immunization center ten years ago. What is wrong with that? By definition, pharmacists shall administer pharmaceuticals as well as deliver them.

I am passionate about this, but some of you may not have the interest in owning your own store for any of a number of reasons. Not the least being money. For most of us, opening a store would mean borrowing the money to get started. Small Business Administration loans go for as 3% interest. Think about it. Don't let fear run your life. Allow anger to guide you.

For those of you who want to continue working in the retail world for a chain, grocery store or big box store, my best advice is to insist that the people who manage you be pharmacists. Express yourself appropriately when there is something that is not working or that you do not like. Communication is vital to your satisfaction in your job. Calm reasoned communication at every point of contention is much better than allowing something to fester and eat you up inside. Remember that anger is good if it is used to guide you. Don't allow it to run you.

The institution will still be the institution, but you do not have to remain institutionalized. Taking baby steps is okay.

# Your Ace of Trump

The laws of pharmacy favor you enjoying a satisfying career. In the end, the laws make you the final arbiter of every single thing regarding the dispensing of prescriptions. Prescriptions cannot be sold if you are not present. Non-pharmacists are not allowed to enter the pharmacy unless you are present and give your permission. There was a time in my career when the pharmacy departments were not locked. (That still is the case in some stores) The entire store could not open unless the pharmacist was present. No matter how you feel about your job, you are the King or the Queen of the store. Pharmacists are the most important employees that any drug store has. Don't forget that. Without you, they can't even call it a drug store.

You are required to counsel patients on new prescriptions. That is the law. If you neglect this, you are ~~an idiot~~ foolish. Counseling is what will differentiate you from the Advanced Certified Technicians who are coming. If you choose not to comply with the law, you may be categorized as less than a pharmacist who is paid $10,000.00 per month. That's the tough love.

The gentle love is that no person can tell you not to counsel and no institution can be designed to make it impossible for you to counsel. The law is the law. If you are mean-spirited, you might actually hope that some Jailer warns you not to counsel so much, criticizes you for the time you take to counsel or simply tells you not to counsel. If that happens, and if it happens consistently, a written complaint to the state board of pharmacy, with complete documentation, is appropriate.

I have portrayed non-pharmacist store managers as villains. Not all of them are difficult. Some of them are friendly managers who are easy to get along with. Some of them know just how important you are. I have worked with terrific store managers. That being said, they are not pharmacists.

*"You don't need a weatherman
to know which way the wind blows"*
*Bob Dylan*

# The Rebels of Comfort.

Pharmacists are well-educated and highly-trained medical professionals, but, in the retail setting, the job of pharmacy is regularly relegated to monitoring The Prescription Mill. This is not practicing pharmacy. It will be independent professional acts that will define what our profession is in the 21st Century. Every independent activity (counseling, primarily) is a profound revolutionary event.

In The Rebels of Comfort, Mister Plagakis presents the difficult situation we find ourselves in, the reasons why pharmacists feel so helpless and what we need to do to get professional pharmacy back in the hands of pharmacists.

## *Price World Pharmacy*

### *Where cost is king*

### *Prescriptions in 15 minutes, guaranteed*

When I refer to *the company* in this book, I will be talking about the mythical *PriceWorld Pharmacy*. *PriceWorld* embodies every evil in our industry that I can think of. The $4.00 Prescription. Free antibiotics and free diabetes medication. The infamous timers that some companies use to motivate pharmacists to work faster. The Drive-Through that relegates the pharmacist to a worker in a certain kind of fast food category. Hair-brain schemes to promote the prescription business such as a guarantee that prescriptions will be delivered in less than 15 minutes.

The executives of *the company* are exclusively non-pharmacists. The CEO is a transplant from a big box store where he had considerable success selling general merchandise, electronics and food and beverages. This CEO had no drug store experience. The executives of *the company* are non-pharmacists and the middle-managers are non-pharmacists with MBA degrees. The pharmacists at *PriceWorld* can be found in the stores or, one level up, as Pharmacy District Managers. The District Managers are supervised by the MBAs.

It was the MBA middle-managers who came up with the ideas of the $4.00 prescription and the guarantee that prescriptions will be ready in 15 minutes or a $5.00 gift card will be handed out. When pharmacists complained, the Pharmacist District Managers acted like the Capos in the concentration camps. They came down with heavy hands to squelch any rebellion because they were afraid of the MBAs. In the end, these unprofessional schemes strip away the patient's perception that a valuable professional service has been rendered. We become cheap.

The model for *PriceWorld* is hierarchical. There is one person at the top of the pyramid and that person is the CEO, the ruler or the dictator. At *PriceWorld*, the CEO is a martinet. He expects the MBAs, the District Managers and the store-level pharmacists to do exactly as they are told. This is not a model that is sustainable for professional behavior. Pharmacists have the right and responsibility to make decisions based on professional discretion. This is not being done at any of the 6,000 *PriceWorld* stores. Pharmacists fear that they will be punished for behaving in a manner that pharmacists should behave.

*The company* is well known for cutting corners in the pharmacy. It paid a multi-million dollar fine for snubbing its nose at the federal government and neglecting to do the proper record-keeping for pseudoephedrine sales. The PBM business it runs has engaged in restriction of trade and has violated antitrust laws. If you Google *the company*, you will find page after page of legal difficulties.

I trust that most of the readers of this book will be pharmacists so I will leave it at that. You know exactly what I am talking about. You could easily add a few paragraphs to the characterization of *the company*.

# The Pharmacist's Role

The role of the pharmacist has transformed during the past six decades and we have not come out ahead. Look at the result, the reality of our situation in the second decade of the 21st Century. The pictures of pharmacists that are presented of us by the APhA predominantly are laughable. The clinician who spends the day doing MTM for a livable wage is one of the fairy tales that comes out of the American Institute of Pharmacy Building in Washington, DC. Those of you who have followed me know that I am one of the legions of veteran pharmacists who have little respect for the APhA. They do not deserve to be able to call the organization the American *Pharmacists* Association because they do nothing to advocate for pharmacists. It is almost as if there is a conspiracy to ruin the profession.

The reality is that most of us do not practice pharmacy at all. We bean-count for *the company*. We tend to the "Prescription Mill". We don't even fill prescriptions anymore. We make sure that they have been filled correctly. Filling prescriptions is no longer a professional task. Well-trained technicians can easily fill prescriptions from the intake to the final product with no pharmacist involved..

Sixty years ago, soon after the Durham-Humphrey Amendment of 1951 changed pharmacy forever, pharmacists had two distinct professional tasks. We filled prescriptions for Pharma-made standardized strengths and dosage forms and we compounded prescriptions.

It took decades before we realized that filling a prescription for thirty tablets or capsules is not a professional task that requires a pharmacist for every single step. Recently we have realized that a pharmacist is not needed at all until the final step and perhaps not even then. Advanced technicians will be the prescription-fillers sooner than later. Advanced Technicians will check the work of other Advanced Technicians. The development and training of Advanced Technicians must be directed by pharmacists and not by the companies.

Filling prescriptions is no longer a professional task that only a pharmacist can do. There are those who argue that filling prescriptions is no longer a professional task at all.

Durham-Humphrey created Big Pharma. Standardized strengths and dosage forms made the pharmaceutical industry. Until recently, compounding has been neglected by pharmacists. It is making a comeback because there is perceived value in non-standardized strengths and dosage forms. Compounding is the quintessential art of the pharmacist. No one else is trained to do it. This is a valuable service.

I am reminded of a story told by a compounding pharmacist in California. The patient presented the prescription and asked how much it would cost. The pharmacist reported that this prescription could be ready the following day and that it would cost $75.00. The patient complained and said that the doctor said that it would cost $20.00. The pharmacist handed back the prescription and told the patient, "Have your doctor mix it then."

Pharmacists who compound must be well paid for their efforts. They have specialized talents that are worth a lot of money. Don't give it away.

The chain drug store pharmacist wears two hats. One is a distinctly professional hat and the other is arguably a non-professional hat. For those of you who hang your hat on running the "Prescription Mill", you are basically just bean-counting. I have watched pharmacists pay practically no attention to what they are doing. They tap the keys with barely a look at the screen. When the computer displays potential problems, they give only perfunctory attention. Running the "Prescription Mill" is not a professional task.

Counseling is a professional task. It is what will define us as a profession for the 21st Century. There are important points about counseling.

Counseling is a legal requirement and if you do not counsel, you are almost hopelessly institutionalized. You say that you do not have enough time, that the demands of the "Prescription Mill" are so heavy that you cannot take the time to *be a pharmacist*. You are short-sighted. What more security could you want? The MBAs at *the company* would be idiots if they told you not to counsel. Counseling is your responsibility and your right as a pharmacist.

I will repeat the advice that I have given loudly and often. Document, document, document. Write down dates, times and who said what. I can't imagine that the pharmacist District Managers for *the company* have a kamikaze death wish. They know that their ass is grass if they dare tell you to break the law.

Professionals do not bean-count. It does not take a Doctor's degree to run the Prescription Mill. Don't blame me. I didn't make it that way. I am just the messenger. A professional is a practitioner who uses discretion and makes independent decisions for the benefit of the patient. Your counseling and my counseling will differ on most drugs. However, neither of us is right or wrong. Both of us behave independently by acting for the benefit of the patient.

The key word is *independent*. *The company* can't tell you how to counsel. That would be idiocy. *The company* can't publish a big book outlining how to counsel on every single drug and every single combination of drugs and conditions. *The company* can't tell you how to counsel a middle-age man who you suspect of being an alcoholic on the use of temazepam. Or how to counsel a pregnant woman on the use of metronidazole for vaginal trich after she tells you that her husband sleeps around.

We spend most of our time at *the company's* stores *working* for *the company* at the Prescription Mill. When we counsel, however, we perform an independent action and that is a profoundly revolutionary act. When you counsel, you no longer work for *the company*. They pay you for practicing pharmacy at their location. They pay you for doing only what you can do, at their location. They can't tell you how to do it. *The company* should be very pleased that you are as accomplished as you are because they look very good when you are very good.

Bean-Counting and running the Prescription Mill are not very satisfying, but they take up so much of your time. It is no wonder that you are so miserable at your job. You are wired and using million-dollar technology, so why doesn't it make you feel any better than five, ten, fifteen, twenty? Counting pills and licking and sticking labels? You spend 90% of your time caught on this hamster wheel.

Pharmacists have been whining about the job for thirty years. Is it any wonder?

Counseling is 100% an independent professional activity. As Forrest Gump would say, "Professional is as Professional does."

I want to be fair. Very few of us can have a 90% Counseling practice and 10% Prescription Mill job because we have not evolved to that yet. Pharmacy is in transition. *The company* has to survive or we would not have a place with thousands of dollars of space and equipment where we can work and practice pharmacy for a very nice wage. We have to do what is needed and wanted to keep ourselves in a good situation.

That being said, you must practice your profession by counseling or they might try to take the opportunity right away from you. You have no choice if you think you will still be able to pull down $10,000 a month and more in the future. Would you pay that much for an employee who only did the work that a well trained Advanced Technician can do for $20.00 an hour? Hell no! *The company* has been studying what you do. They would love to pay you less.

# Point of Law

There is a law in every single state that every one of us must be willing to go to the barricades to see to it that this law remains in the books forever. We must be willing to temporarily sacrifice our time, our money, even our well-being to protect us from forces that are hell bent on eliminating our profession eventually. We must be willing to get dirty and bruised and bloody if any force tries to change this law. We must aggressively enlist the public for their support. Look out in Arizona.

The law that I am speaking of is this: *There must be a pharmacist present when a prescription is dispensed to the patient.*

The law makes sense. Counseling needs to take place. The patient may have questions. The Boards of Pharmacy are mandated to regulate our profession for the benefit of the public. It is a no-brainer that the health and welfare of the public could be endangered if they eliminated this law. Look out. The boards have not shown consistently that they are good thinkers.

*The company* doesn't care about the patient no matter what they say. The MBAs could not care less if Maggie Jones dies because she was not properly counseled. *The company* would love to fill the prescriptions in a remote location and deliver them to a dispensary store and have the prescriptions sold by a clerk. You can help prevent this nonsense by showing your value as a professional. Trust me! You get no professional points for running "The Prescription Mill".

# I'M TOO GOOD LOOKING TO BE THIS OLD

The good news is that there is going to be a hell of a lot of money to be made from selling drugs for many years to come. As long as you have to be there when it is sold, by law, you should be fine.

In 1999, Americans spent $104 Billion dollars on prescription drugs. In just nine years, that amount grew by $130 billion dollars to a whopping $234 billion dollars in 2008. That is an increase of 128% in less than a decade. The baby boomers were still not official in 2008.

Since January, 2011, 10,000 Americans will turn 65 years old every single day. Many of them are not going to give up their "youth" no matter what and they will have the money to keep themselves vital, healthy and definitely sexy for the entire ride.

They will buy the time of personal trainers and they will practice yoga. They will buy specialized healthy food. They will flood physical therapy gymnasiums. They will seek out Botox Clinics and they will buy the drugs that Big Pharma will provide for their needs. And, make no mistake, Big Pharma will give them everything they want. Viagra was the first. Pharma has put the mission of ameliorating diseases and illnesses on the back burner. There will be more money in "Boomer Medications".

The Boomers will challenge us and there are still the other 200+ million Americans who are not that mature yet. They get sick too. How long will it take until our industry sells $1 trillion dollars worth of prescriptions?

As long as we keep "The Law", we will be really busy. The Prescription Mill will be run by Advanced Technicians and pharmacists will be spending most of their time practicing pharmacy. Immunizations included. Counseling. MTM. Compounding and whatever else the alphabet soup organizations (APhA and other organizations) come up with.

My advice is that you start now. Show some respect for yourself and your education, your colleagues and the family that supports you and believes in that Doctor of Pharmacy degree, even if you have lost faith. You are an intelligent, highly-educated, well-trained medical professional. Start acting like it. If you don't, you will be violating both federal and state pharmacy laws. *The company* will not support you if you are cited by the Board of Pharmacy. An MBA at *the company* will issue a statement that will read like this:

"All of *the company's* pharmacists are very well aware of our policy. It is set in cement. There is no wiggle room. Our pharmacists are expected to follow all laws and rules regarding the practice of pharmacy, both federal and state. It was Mister Johnson's personal decision to violate the law by not properly counseling Ms. Olive Peterson. The damage that Ms. Peterson suffered is the sole responsibility of Mister Johnson."

It is interesting that *the company* wants you to be a lemming until something bad happens. Then *the company* doesn't just lead you to the cliff, they push you over the edge, all alone.

# All Hail the Emperor

In 2007, the editors of *Executive Profile Journal,* an important monthly among business executives, published a psychological profile of the Chief Executive Officer who received the biggest bonus in the retail drug store industry. That person was the CEO of *the company,* Mister George Fullovit.

Accompanying the report compiled by Dr. Jerrold Porter, a psychiatrist who directed the business executive psychology program at George Sutherland University, was a photo gallery of Mister Fullovit's estate "Fresh Breezes" on the Gold Coast of Long Island. Displayed was Mister Fullovit's stable of classic cars, including a fully restored Cord Roadster and a pink Cadillac that had been owned by Elvis Presley.

It must be noted that both the Malibu beach house and the 8500 square foot Lake Tahoe Sierra Nevada mountain estate that were shown in the article are now the property of Mrs. Sean Nyland, Mister Fullovit's ex-wife. Mrs. Nyland is now the wife of Fred Nyland, the CEO of Public Area Stores, a large grocery retailer in the southern United States.

*Executive Profile Journal* is a small publication that prides itself on being independent. Ostensibly, it was reported that *the company* was ready to take legal action against the *Journal*. However, it was well know in the retail drug store industry that Mister Fullovit was the only reason why the legal staff of *the company* pursued the ill-advised legal steps that cost *the company* over $100 million in a losing battle. Doctor Jerrold Porter filed a counter-suit and eventually settled for $10 million.

Doctor Porter and his staff of psychologists who specialize in the dynamics of big business published the opinion that George Fullovit was a delusional narcissist who believed that *the company* was in a war with the enemies being the other retail drug store businesses. They also stated the opinion that he was an impulsive showman who would take all of the credit for successes, but would not hesitate to pass the blame on to the inhabitants of the MBA executive suite for failures. There was an in depth analysis of the ill-advised *15 Minute Prescription Guarantee Program* that was deemed illegal by the state boards of pharmacy in all fifty states. It was well known that Mr. Fullovit backed the program, but retreated when it failed. Four MBA executives lost their jobs.

The consensus was that Mister Fullovit acted like a sociopath. The executives of the other large retail drug store chains did not like him. They understood that they were in a culture that needed a certain level of cooperation. Mister Fullovit stood alone and consequently *the company* stood alone. The study's psychologists were the foremost experts in the study of business executives. They speculated that Mister Fullovit was a coldly calculating strategist and crazy, like a fox.

Doctor Porter said, "Mister Fullovit is a leader who dominates the company and can act virtually without constraint. Among the MBA executives are a number of sycophants who have benefitted from their unexamined agreement with anything and everything that Mister Fullovit wants to do." His megomaniacal self confidence has permitted him to make bold initiatives and, in the management culture of *the company,* very few MBA executives who object to his initiatives survive.

The science behind the studies that the Journal publishes is a clinical-case approach. They also ground their studies far more firmly on biographical facts than on Freudian speculation or personal opinion. A software program designed by researchers at George Sutherland University found an alarming frequency of words like "I". "Me", "Mine" in Mister Fullovit's writings, interviews and speeches. There was an interesting lack of words like "Us", "We", "Our" signifying that Mister Fullovit's proclivity is to go it alone. Mister Fullovit's well-used management meeting phrase, "We are gonna wipe them out." was repeated in numerous speeches and reflects very high power orientation.

A researcher by the name of Ronald Benson, a PhD psychologist at Auburn University who participated in the study, stated at a news conference that, while usually rational, Mister Fullovit can be prone to delusional thinking when under stress. Doctor Benson reminded the reporters of the billion dollar losses that *the company* suffered when they were ordered by the courts to break up the company and sell off their huge PBM business after being convicted of violations of anti-trust laws. The losses could have been minimal, but Mister Fullovit refused to listen to the lawyers. It should be noted here that three of the MBA executives from *the company* who lied under oath are now in federal prisons, doing ten year sentences, for lying to the grand jury and for obstruction of justice.

At a news conference after he had been accosted by two thugs in a parking lot, Doctor Benson shook his head and said, "Fullovit is a throwback to the 19th Century. Modern companies are not run this way. They do not fully utilize all of the talent who work for them. Some of the best minds at *the company* are the pharmacists in the stores. They are smart enough to have earned Doctor Degrees. They are retail pharmacy specialists and they are treated no better than clerks." Doctor Benson was asked if he had a prejudice against Mister Fullovit. He laughed, touched his bruised cheek and said, "Yes, I do." The reporter asked if Doctor Benson believed that the parking lot incident had anything to do with him participating in the study. He touched his black and blue cheek again, smiled and said, "Yes, I do."

# Leadership and Power

These are distinctly different. Leadership is when an executive starts a parade and everyone follows him because it is intrinsically a good idea. There is no coercion. No force or manipulation. The pharmacists in the stores say, "Yes, I can do this. It is a professional task and I can even get some money from it."

One of the major drug store chains is urging their pharmacists to deliver M.T.M. services. Insurance companies pay the chain because they anticipate superior outcomes. This chain pays the pharmacist for the delivery of M.T.M. The amount is, at this time, a token, but it is a start. Eventually, drug store companies will see the economic value of urging the pharmacists to practice pharmacy. Eventually, pharmacists will be highly rewarded because the insurance companies will be saving a great deal of money. Of course, professional behavior will be rewarded because services such as M.T.M. will reduce waste, increase compliance and help to reduce medical costs in this country.

I spoke with a pharmacist at this chain who proves M.T.M. services. "I do it on my weekend to work. Consistently, the patient is surprised when I call. Most of these people take multiple drugs and are pretty sick. They seem to feel that someone actually cares about them".

Power on the other hand is a much more ambiguous *line of attack* and I use that phrase on purpose. The hierarchy of most companies that have retail pharmacy businesses have a *dominator* and the *dominated.* The pharmacists in the stores, of course, are among the second group. It does not matter how ridiculously ill-advised an initiative is, if the *dominator* throws his considerable weight into it, thousands of people are expected to toe the mark and make it work no matter how impossible or ridiculous it may be.

The C.E.O. of most companies rely on power to get what they want. Take George Fullovit, for example. Mister Fullovit sees his position as all-mighty. He is the dictator and he acts like it. He believes that all power flows from him and that it flows only one way. The idea that power could flow from the *dominated* upward is laughable to Fullovit.

The middle manager MBA executives only have power because they have access to the dictator. They are powerful because they can communicate with the all-powerful emperor. These are the police who wade into the crowd with Billy clubs to enforce the edicts of the dictator, Mister Fullovit, in our example. They may be called Regional Managers or District Managers, but don't kid yourself. Their job is to make sure that you do what Fullovit wants you to do, his megomaniacal vision being the only thing that counts.

Fullovit and the MBA executives are seminarians in the religion of domination. They truly believe and stake their lives on the idea that all power flows from the top down. They are wrong. Power flows both ways. Power flows both ways in all power relationships. All power relationships are interactive and reciprocal. In truth, the rule of a dictator and his elite group is always weak and unstable because it is always dependent on the cooperation of the *dominated.*

**SAME GREAT TASTE
FRESH NEW LOOK**

RITE AID
QUALITY SEAL
CIGARETTES

RITE
AID

Recently, *the company* induced a 15 minute guarantee for three prescriptions. If the prescriptions were not ready in 15 minutes, the patient received a $5.00 gift card. Thousands of gift cards were distributed. There were some exceptions. The 15 minute guarantee did not apply when the doctor had to be consulted or when there were insurance problems. It did not apply when professional services were required. The pharmacists decided that the professional service of counseling was required on all new prescriptions. Game over! The emperor wore no clothes.

Mister Fullovit signed off on this program because he believed that it would pull *the company* out of a downward spiral. His executives, the MBA sycophants, with fawning attention, assured Mister Fullovit of "What a good boy are you." *The company* put millions of dollars into the 15 minute guarantee. They cajoled and intimidated pharmacists in the stores to no avail. *The company* had not counted on the patients understanding the importance of being well-informed about their medications and the drug therapy.

The religion and science of the worst companies in the drug store industry is all about domination and control of the people who actually do the work. The power of Mister Fullovit is voracious. This kind of power takes everything that is there to take. Mister Fullovit makes no distinctions between right and wrong because he is always right. He knows that he must expand his power or die. His goal is to fill all vacuums with his own will. His method is to crush the weak and this shows by the number of non-pharmacist MBA executives he has forced out of *the company*.

George Fullovit has been corrupted by his lust for power. He fully expected the pharmacists (dominated) to believe that he was committed to helping them. He presented the idea that the 15 minute guarantee would make *the company* a better place to work. He actually sent out memos that he was doing the will of the pharmacists in the stores. Mister Fullovit was in love with power. Power for power's sake. It was like an opiate. The more he got, the more he needed.

What Fullovit wanted was only to acquire more power and since power emanates from below, he knew that he had to have the cooperation of the dominated. The pharmacists did not go along with Mister Fullovit's program. As a group, the dominated refused to cooperate and, since all power really emanates from those below who actually have to carry out any program, the 15 minute prescription guarantee failed miserably. *The company* was the laughingstock of the entire drug store industry. The dominated refused to cooperate, the pharmacist district managers gave the program lip service and hunkered down just to survive. Without the cooperation of the pharmacists, no power was sent upward to the executive suite. Power is the lifeblood of any plan or program, good or bad. In this case, the dominated refused to cooperate. There was no power returned upward.

This was a fortunate event in the history of the profession of pharmacy. The 15 minute guarantee was highly publicized. The attitudes of pharmacists all over the country were noted. Pharmacists who did not work for *the company* engaged in a campaign to get the boards of pharmacy involved. Official organizations like the Institute for Safe Medication Practices came out publically and took off the gloves in their criticism of *the company*.

Pharmacy benefits from the failure of *the company* because the failure was so dramatic and so well publicized that it is doubtful that any other drug store corporation will ever dare try it again. Pharmacists all over the county became involved. Concerned citizen groups petitioned the state boards to outlaw such programs. Pharmacists with *the company* decided to practice pharmacy. They liked it. It made them feel good. When they were officially reprimanded for their independent professional acts, the state boards stepped in. This was not only an embarrassment for *the company*, the fines that were levied were substantial.

*The company* still seems to live in the 1970s when delivering a commodity was the only job a pharmacist had. Delivering the prescription fast and cheap was the goal that every company aspired to. Without fast and cheap, there was no competitive edge to be gained. It must be noted that Mister Fullovit was the CEO of a big box store company that was the first to promote fast and cheap prescriptions. That was in 1971, however. It is 40 years later and the model for a successful pharmacy operation includes providing professional services meant to provide health care that can enhance the benefits of drug therapy.

Without at least the passive support of the pharmacists at store level, the executives become just another bunch of crackpots whose ideas very likely are lousy. I worked for a Washington State corporation that went through three CEOs in rapid succession. The middle CEO was a Los Angeles businessman who hung his hat on cheap imports. He reconfigured every single store into a stadium set up where the drug store was on the edges and the middle of the store was a fast turnover import junk shop (my characterization). The prices were terrific, but we, at the store level, knew that a relatively sophisticated Seattle and Western Washington clientele did not want $2.99 sewing sets or $3.99 clusters of cheap plastic kitchen ware. We knew that we wouldn't buy this rubbish and we knew that our average customer wouldn't want it. They hated it so much that they started to go elsewhere for their prescriptions.

To make matters worse, the CEO instituted a dollars off coupon program for prescriptions. For every $50.00 spent on middle of the store merchandise the customer got a coupon. I think it was worth $2.00 off on any prescription. This was in the late 1980s when the majority of prescriptions were still cash and the average prescription was less than $20.00. Pharmacists hated this program.

We felt that it cheapened our operations. We complained mightily to our Pharmacy District Managers, but it is impossible to get people whose wage is dependent on *not agreeing* to *agree* with us on anything. At least not on the record.

I remember a conversation with my Pharmacy District Manager. "Do you think I like this, Plagakis?" He stabbed his finger at me. "You would be smart to just keep your mouth shut and fill prescriptions. Get your reports in on time and go home to your family at night." He frowned. "Every one of you pharmacy managers gives me grief. There is nothing I can do about it." Then he smiled. "At least not officially."

The CEO had no idea that customers were laughing at us. He didn't know that every single pharmacist in the company called him *Porter the Importer* when his name was Vardabarian (my best recollection). His grand scheme failed miserably and it was said that he had no clue why.

Hierarchical organizations, like every drug store company that I know about, require the cooperation of people at every level to be successful. That is why it is imperative that the person at the top implements programs that everyone can conscientiously support. My best advice to any CEO is this: *Once they start laughing at you, you are doomed.* When enough people withdraw their support, for a long enough time, the power of the CEO disintegrates.

I am not painting with a broad brush. There are plenty of benevolent CEOs who make decisions that store level pharmacists benefit from professionally. The fact remains, however, that while the corporation appears to be permanent, it is subject to the same weaknesses and vulnerabilities as the human beings who work in the stores.

The corporation wields enormous power over our lives. It controls how we spend our time at the Prescription Mill. It controls when we take our vacations. It makes demands in every area except our professional discretion. We are not piece workers in a factory. We are medical professionals and the law, personal standards and ethics often do not agree with the short-sighted goals of fast and cheap.

Your drug store company is not made of steel. Even the biggest pharmacy business is nothing more than a community of ordinary pharmacy workers (Pharmacists and Technicians). Ordinary workers can be stupid, irrational and even incompetent. Many of them hunker down in their slot in the hierarchy and are jealous of their bosses and look down at their "inferiors". Ordinary pharmacists can also be heroic, stubborn, non-conformist and moral. These are the ones who are not afraid to speak up and stand up for what they believe is right.

A monolithic, hierarchical model is the conception of power that those at the top like to perpetuate. However, this model of doing business depends on everyone following the same "rules". But it will only works if both the rulers and the ruled believe in the same "policies". The CEO of most drug store companies makes store visits. They come back to talk with the pharmacist. The CEO asks questions and the pharmacist often just blows smoke.

"Yes sir, Mister Fullovit, the new program is showing results." But the pharmacist is thinking: *"Where the hell do you think we are going to find the time, Bozo?"*

Unfortunately, the leaders in *the company* often try to portray themselves as god-like. The want to look like that know everything and can never make a mistake. But managers, including the MBAs, are really just ordinary people too. In the 21$^{st}$ Century, they make decisions based on gut feeling. Look at what they spend money on. Much of the technology is not functional. It is just ostentatious and showy. To justify the purchase, they expect the pharmacist to use it even though it is a time-waster. Many of you are in-store pharmacists and have had the opportunity to be forced to listen to incompetent *experts* simply because someone hired them.

I remember a fellow showing up at the pharmacy I managed with credentials from the corporation. He was an *efficiency expert*. He was hired by the corporation to assist the pharmacy managers in running a more efficient business. He came in with binders and a big smile. He called me *James* and spent two days spying on us and then presented his findings the third day. He told me that it would be more efficient use of the pharmacy staff if I assigned a technician to take telephone prescriptions. I listened and, of course, disregarded that advice and everything else he advised. I trust that the corporation paid this guy a sizeable fee to waste our time.

I suppose it would be best if all pharmacists said a daily prayer for the CEO because he/she is just an ordinary person with all of the flaws, blemishes and defects of any other ordinary person. It is pure luck if the CEOs business moves work. Good or bad, his decisions depend on the agreement of the pharmacists in the stores. Without you, the corporation will bumble and stumble along. *The company's* disastrous 15 minute prescription guarantee shows what can happen when force is the only way they can get the pharmacists to cooperate.

The only thing that sustains the illusion that *the company* is a major player in the drug store industry is the deception caused by the size of the company. That and the silent support the workers give the company because of the fear of sanctions. I have done things in my career that I would have never done had I not been afraid of sanctions. I was afraid of the *write-up*. I didn't want my employee file to have anything negative in it. I was afraid that I would be denied a summer vacation and be forced to take my vacation in October or March, if I did not toe the line. I was afraid of losing my job. I had a car payment, a mortgage and my daughter had voice lessons and was a black belt in Taekwondo.

I didn't realize then that the corporation was not omnipotent, like a religion. It was not invulnerable. The CEO was an emperor with no clothes. There wasn't even a palace insurrection. When the corporation stalled out because the people in the stores gave only lip service to the initiatives that came from the executive suite, the CEO took his golden parachute and disappeared to Costa Rica, I heard.

My characterization of George Fullovit is a design meant to illustrate the damage that can be done when the person at the top of the pyramid in a drug store company does not understand where the real power resides. The essential fuel that runs a company comes from the bottom. Mister Fullovit does not understand this simple fact. He wants the pharmacists to believe that he is committed to helping them. In fact, his only goal is to get more power and he may not even know it.

# Obedience

If some of the programs that come down from above are so ridiculous, why do we obey and bust our backs trying to make a failed policy work? Remember, without the support of the pharmacists in the stores, the most powerful dictator in the drug store industry becomes just another crackpot with ideas that are doomed to failure.

We obey out of **habit.** This is the way it is and it has always been this way. We do not question what our superiors are asking of us because that would be too difficult.

**Fear of penalties.** It is the fear of penalties, not the penalties themselves, that is the most effective method of enforcing obedience.

Some of us feel that we have a **principled responsibility** to the company to do our best no matter how ridiculous or unprofessional their initiatives are. It is almost like a religion.

We may obey for **selfish reasons.** We can get monetary bonuses and we can get recognition and enhanced prestige.

There are pharmacists who **identify with the dictator.** We can get incredibly high when something works and incredibly low when it fails.

Then there are the pharmacists who just **don't care.** They are indifferent and obey without question. *The company* had pharmacists who did everything they could to satisfy the 15 minute guarantee. They neglected their patients, but they obeyed.

There are many pharmacists who have **a lack of self-confidence** and hand control of their lives over to the ruler. They feel inadequate and need to be governed.

No wonder so many of you are so miserable at your job.

# The Great Motivator

Fear is instilled in us at an early age. It affects our perceptions throughout our lives. For me, the world was a very dangerous place, according to my mother. The dangerous things I did were behind her back and I was still scared. From an early age, she was all over me to, "Wear your rubbers, Jimmy. Wear your rubbers." She wasn't talking about condoms. She was making me wear overshoes. Had she said, "Wear your condoms, Jimmy. Wear your condoms" I would have slept better many nights during my senior year of high school... In 1958, condoms were kept in a drawer in the pharmacy. Buying them made me feel like a pervert.

I was the only one of my friends who had to wear his rubbers. I felt like I wasn't as good as my friends. My mother taught me to be afraid of just about everything. Fear was a primary motivator in my life. The essential result was that I didn't feel very good about myself. I brought much of that to my career as an employee pharmacist.

George Fullovit is a master at using fear to manipulate and control his pharmacists. His store managers have been tutored on how to use the dreaded *write-up*. I had a pharmacist complain to me that his wait time was thirty minutes and that he had been written up twice.

"I don't have time for transfers" he complained. "I have a family to support."

"Do you really believe that you will lose your job because of your wait time?"

"I can't take that chance." It was fear that controlled his life at work. A man this fearful most likely is dominated by unexamined demons.

You are afraid of crime so you work only in "safe" stores, in the suburbs.

You are single and you are afraid of AIDS so you don't have sex even if you are really attracted to that person from the book club.

You are so afraid of immigrants that you tell a man with dark skin wearing a turban that you don't carry losartan. You just want him to leave and never come back. You find out that he is a doctor and excuse your behavior by making the excuse that even doctors can be terrorists.

You are terrified of making a mistake. You work too fast and get distracted easily. You wake up and night reviewing prescriptions and suddenly think that Mrs. De Marco got the wrong drug.

You are frozen by fear that you could lose your license. *The company* sent out a memo. Company policy is to triple-check every prescription. How is that possible? Who has the time? You are not stupid. You know that *the company* will leave you twisting in the wind if there is legal trouble.

You are afraid that your District Manager will send you to a store 60 miles from your home if you don't behave yourself. The life of a floater is precarious.

You are so afraid of failure that you never take chances. You have been passed over for promotions and you wonder why.

When it is time for your annual job performance assessment, you expect the worse. That fear of losing your job is always lingering in your mind. *The company* is very skillful at planting doubts. The MBAs have the idea that employees who are worried will work harder for their security, including working extra, off the clock.

You really want that vacation in July this year. You are afraid they won't give it to you so you work extra shifts to cover other pharmacist vacations just to look good. You follow every rule. Your spouse has the time off and there is a family reunion.

You are always afraid that you will have to miss your daughter's plays or your son's games because you have not earned the time off.

Fear limits our freedom and creativity. It prevents us from thriving because fear will not allow us to takes chances. You are just plain afraid. You learned when you were young that you better obey. You obey *the company* without examining why.

*The company* takes the place of a church for many of us. We follow *the company's* rules to the letter because, if we don't, we'll go to some kind of drug store hell.

Fear can be the reason we do things and it can be the reason we don't. We come to work early because we want the pharmacy to be ready for the morning rush. We are afraid of what the store manager will say when there is a mob at the register.

We don't complain about not having a technician for the first hour because we are afraid of being accused of not being a team player. *The company* rewards team players. The pharmacist who wins "Team

Player of the Year" gets a Hawaii vacation for two. You have met the winner from last year. He had a nervous breakdown a few weeks after he and his wife returned from Honolulu. He pulled down his pants and mooned Mr. Fullovit when the CEO showed up with a cheery smile and, "Terrific job, Jimmy." The pharmacist's name is Jerry.

When you are fearful, it is impossible to not be totally oblivious of the power that a pharmacist has. You have immense power. You are not a clerk in the camera department. You are a medical professional with the discretion to engage in professional acts any time you feel it is appropriate.

When George Fullovit took over as CEO of *the company*, it was a modestly successful drug store chain. The employees were happy, but Mister Fullovit wanted to see motivated associates. He liked to tell the MBAs, "They are either with me or against me in this turn around. Fire the non-performers. I want to see that dull look of fear in their eyes when I visit the stores." That is when abject fear took over at successful stores. Those formerly happy pharmacists started to look over their shoulders.

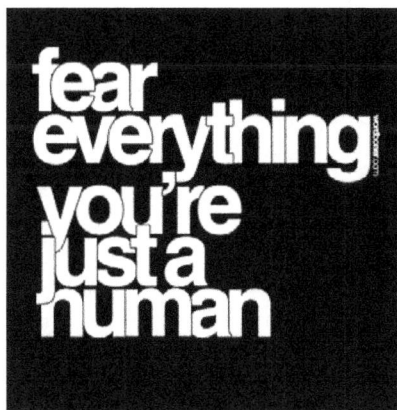

**fear everything you're just a human**

# They Became Fearful Of Penalties

The penalty that is most prevalent on your mind is losing your job. You have a spouse and kids who attend a Montessori school. The mortgage on the new house is hefty and you still have student loans you are paying off. You have to keep this job. The MBAs at *the company* know that you are fearful of losing your job. They are adept at playing chicken with pharmacists. They usually win because it is the pharmacist who blinks.

Brutal penalties make us wonder if George Fullovit and his MBAs might just be sadists. This is not a benign way to control the professional staff of any company. It is mean and derisive. It can cause real damage. GERD, hypertension, insomnia, sexual difficulties, spousal abuse and even suicide have been noted among pharmacists at *the company* and at other drug store corporations. It is a pathetic commentary on George Fullovit and his methods. He is not a good leader. Mister Fullovit is a brute. He uses force and coercion to get what he wants. He wants talented, well-educated medical professionals to act like robots. Often, he gets what he wants because the pharmacist is too scared to do anything else but obey.

No wonder you can feel so hopeless. The situation that so many pharmacists find themselves in is vicious. It is almost violent. Of course, we aren't laborers who just might bang the foreman over the head with their lunch pail. We are professionals who quietly accept the violence laid on us. At least I think we are still professionals.

An acquaintance who works for *the company* in a small town in a southern state is well respected in the community. Customers call him "Doc". *The company* is very fortunate to have him.

His District Manager, following Mister Fullovit's orders to "Make them jump like lobsters in boiling water," took this Pharmacy Manager aside and told him that he was just a count, pour, lick and stick pharmacist. The DM told him that he would have to start giving the patients better service. In the next sentence he told him that he had to fill all prescriptions, up to a batch of three, in 15 minutes or he could be written up.

He then announced that a beer cooler would be installed next to the drive-through window. The pharmacy staff would be expected to sell six packs. A Frito-Lay chip display would be next. Deli sandwiches were being tested in select stores. Cigarettes were being considered. "It's about the bottom line," he said, "This is a resort area. There are vacationers who will get in line so they don't have to get out of their cars." He looked down at his fingernails and muttered, "I'm surprised you are not complaining and that is good." He grinned. "Very good. This is the new, modern version of *the company*".

The District Manager rolled the dice and they came up snake eyes. This pharmacist had been with *the company* when it was a happy place to work. He knew the difference. He made eye contact and said, "So sue me."

"Sue you? Sue you for what?" the District Manager said. His eyes looked suddenly rheumy.

"You better do it because I sure as hell will sue you and this pathetic company."

"You can't sue us?" Our pharmacist is a big man. The District Manager rather small. He stepped back two steps. "What will you sue *the company* for?"

"Leave me alone or you will find out in court," the pharmacist warned. "Your career with *the company* will be over." He smiled. "No beer. No sandwiches."

Some of that story is true and some of it is made up. It is an illustration of how the threats of penalties can diminish the power of those who use them. Usually, the use of threats only gets temporary obedience. Corporate initiatives die a sudden death when the employees with their feet on the floor do not give their support.

Threats of penalties do not create power. *The company* turned to penalties because George Fullovit and his MBAs felt their power slipping away.

The most violent penalty of all is to take away a pharmacist's job because he or she does not live up to the expectations of his or her managers. They fire people for many reasons, but when they do it because a good pharmacist refuses to engage in dangerous or unprofessional conduct, the firing is especially violent. A hundred years ago, they sent in labor goons with clubs. In the 21st Century, they build a case and then they fire you. Usually, the case is as flimsy as showing that you did not follow company policy even if the policy is bad. It is still company policy.

**A parenthetical note: We must always remember that we are professionals and that there are professional ethics and personal professional standards that should always trump company policy. The aces of trump, of course, are state and federal pharmacy laws. Practicing pharmacy is not running the Prescription Mill the way they want you to. Practicing pharmacy is all of those independent professional acts that require discretion. Discretion is what defines a professional.**

I will return to *the company's* 15 minute prescription guarantee. This is an insult to competent pharmacists. It relegates professional employees to the role of assembly line workers in a factory. It is unrealistic to expect pharmacists to do everything that they feel is necessary to serve the needs of the patient in 15 minutes. To penalize pharmacists for not toeing the mark when it is clearly not in the interests of the patients for the pharmacist to be rushed and distracted is immoral.

Violent sanctions that are meant to control pharmacists are draconian. When *the company* used its superior force (they wrote the pay checks) against pharmacists, they found out that they did not get the desired result. It was up to the Pharmacy District Managers to do the dirty work. They became unreliable. They made a good show of obeying, but they were reluctant to punish pharmacists who were doing nothing wrong. The DMs, who were pharmacists themselves, began to sympathize with the pharmacists in the stores. They felt that George Fullovit and his cadre of MBAs were losing their legitimacy as leaders.

It is the threat of brutal punishments that create the fear that controls people. George Fullovit eventually began to see his grip slip when he was thrown off balance and weakened considerably by being too brutal. He thought that he would be setting an example when he ordered the firing of a number of Pharmacy Managers because they disregarded the 15 minute prescription policy. What he got was the famous "Fifteen Minute Five". The five pharmacists he fired became martyrs. Customers complained. Five pharmacies that had been well run became quagmires when the inexperienced replacement managers tried, and failed, to live up to the 15 minute rule.

Pharmacy Managers who had been neutral were appalled by the firing of loyal employees and they became enemies of Fullovit and his MBAs. Fullovit had, in effect, put a pistol to the heads of the pharmacists. When he started pulling the trigger, he lost control completely. MBAs began to find other jobs and resign. The drug store industry labeled Fullovit a renegade. The Board of Directors of *the company* started to draft a future that did not include George Fullovit.

**The only power that any leaders have is the power that is given to them by the people they govern. The executive who does not understand this is doomed to failure. Fullovit thought that he could force pharmacists to engage in unprofessional behavior. He was wrong.**

I do not like to talk about winning and losing. I prefer to think of our situation with our employers to be an opportunity for a win-win result.

The certain way to fail to achieve this result is by unionizing when the drug store company you work for is not a bully. That would be the easy way to proceed, but I don't believe that the outcome would be the best. Pushing for unionization would be a confrontational process rather than a cooperative process. Enemies would be made. Large corporations are experts at fighting off threats and the smell of a union is a huge threat.

It may seem counter-intuitive that non-confrontational measures can be effective in gaining what we want in companies such as *PriceWorld.* Mister Fullovit and the MBAs at *the company* have substantial amounts of force at their disposal. They know only one method to proceed. They will always defer to bullying you to get you to implement the programs that were designed in the executive suite. At *the company,* unionization may be the only viable answer.

One of the essential problems that many CEOs have is what I will call the *emperor syndrome.* They actually believe that they can do no wrong. They believe that the ideas that they sign off on are all good ideas, workable ideas and are business plans that are bound to be successful. We know that this is rarely the case. Plans and programs from the executive suite are usually ponderous and as hard to move as a stalled Mack truck,

The best ideas always come from below. Perhaps because there are a lot more people down there and all of them are smart and well-educated. Many of them have had the life beaten right out of them by *the company.* However, there are plenty in intellectual hibernation waiting their chance to contribute. They are well-experienced and have the feet on the floor knowledge that the MBAs do not have. The most effective CEOs do not like being all alone at the top. They seek out the pharmacists who are doing the heavy lifting and who have the good ideas.

The name of the business is *pharmacy* and *pharmacists* are the only people who are legally allowed to practice pharmacy. The best way to prevail as professionals is when the hearts and minds in the executive suite at *the company* come over to the side of the pharmacist and begin to encourage the pharmacists to practice pharmacy first, in all cases. I know. I am falling down laughing too.

The best CEO was the owner of the company. He had smaller free-standing stores all over California. He listened to ideas from the pharmacist store managers at every chance. His job as CEO was to travel from store to store to listen. He took the store managers for meals. He asked questions and he listened.

I managed the drug store as if I was the owner and the CEO encouraged me at every turn. The CEO financed my partnering with a CPA and a software programmer to develop a pharmacy computer system. This was around 1972 when there was no such a thing in any pharmacy anywhere. It never panned out because the CEO's pockets were not deep enough. However, he owned the software at the end and he had a very motivated manager in me.

The store was a money-maker right up until I quit for a five year adventure. I do not remember this CEO ever implementing his own ideas. He depended on the pharmacists in the stores (Managers and Staff) to give him guidance. When I wanted to send out a big advertisement in the mail, at considerable expense, he didn't hesitate. When he did not hear from his pharmacists, he would call and ask if we needed anything. Had he tried to control us, he would have failed. Had he insisted that his way was the only way, he would have failed dismally. In the end, he made a lot of money by simply respecting his most important employees. He and his wife lived a good life in Tarzana, California, in a house on a hill. His pharmacists were happy and loyal. This man built relationships. There was mutual trust. The human element. Without this, there is nothing.

I believe that the best we can expect is when the minions up above continue to believe that they are infallible, but realize that to fight the pharmacists in the stores is too costly. The 15 minute guarantee at *PriceWorld* was just another insult upon previous insults. There was a wide opening for Fullovit to back off, lick his wounds and cut his losses. The enraged pharmacists did not see the opportunity. They were misused and angry and they tried to fight force with force of their own. They called upon the Guild for Professional Pharmacists (A union) and started to organize. This was playing into Fullovit's hand. They handed him the Joker of trump. He was an expert in fighting unions. The pharmacists kept him alive. I can't imagine any drug store company welcoming a pharmacist's union.

**Confrontation is not the solution for getting the control of the practice of pharmacy back in the hands of pharmacists. Separating the Prescription Mill from the practice of pharmacy can be the answer. No executive in their right mind is going to tell you to not counsel, for example. When you use your discretion to engage in an independent professional act, you are a rebel regardless if shots are fired or not.**

I do not believe it is realistic to expect corporations that have millions of dollars invested to give up control of the Prescription Mill. Would you? We should never threaten the authority of the corporation when the Prescription Mill is in question. We should manage the Mill effectively and efficiently until the laws that allow Advanced Technicians to implement *Tech check Tech* are in place. There must be a balance, however. You must practice pharmacy. It won't be easy. You said that you wanted to work when you asked for the job.

The words *power, confrontation, violence and fear* are central to our discussion of the dynamic that pharmacists work within. It seems to me that it works both ways. Mister Fullovit and his MBAs have to be terrified. They are all alone up there and they have to depend on people who do not like them to save their asses.

When George Fullovit and his MBAs use harsh tactics their power is diminished. Power and sanctions like the "dreaded write-up" are not the same. Power is a psychological, moral force that makes people want to obey. Sanctions such as *do it or else* enforce obedience through coercion well spiced with fear.

Those who use fear and coercion may manage to impose their will, but their control is always tenuous because when the coercion ends, or the threats lessen, there is even less incentive for the pharmacists with their feet on the floor to obey. They have enough to do. If nobody is watching, why bust their butts to facilitate a failed program.

Force is the way of the impotent. Those who are effete managers with no power have to rely on force to get what they want. Force does not create power. People respect power. They laugh at force. You were all young once. You remember what happened when they weren't watching you.

When George Fullovit tried to use force to implement the 15 minute prescription guarantee it was because he feared that his ability to influence events was slipping away. Corporations that try to rule through force are weak. Weak CEOs have always tried to rely on force to compensate for their powerlessness.

**The executives of every drug store company need to pay close and careful attention when they implement a program that requires the cooperation of the pharmacists with their feet on the floor. They need to seek the advice and *consent* of the people who will be doing the work. Without this, money, time and effort will go into grandiose plans that will go to seed without any bang for the buck. *The company's* 15 minute guarantee was a joke and George Fullovit was the laughingstock of the industry.**

**JOIN THE REVOLUTION**

*You say you want a revolution*
*Well, you know*
*We all want to change the world*
*You tell me that it's evolution*
*Well, you know*
*We all want to change the world*
*But when you talk about destruction*
*Don't you know that you can count me out?*

*Don't you know it's gonna be*
*Alright?*
*Alright?*
*Alright?*

*You say you've got a real solution*
*Well, you know*
*We'd all love to see the plan*
*You ask me for a contribution*
*Well, you know*
*We are doing what we can*

*But if you want money for people with minds that hate*
*All I can tell is, brother, you'll have to wait*

## John Lennon and Paul McCartney

The revolution does not have to be confrontational. Some of you are very angry and are aching for a fight. I contend that the need to raise a clenched fist is not helpful and is non-productive. I believe that trying to destroy a monolithic organization such as a drug store company is not realistic. The company is people, not buildings and computers. The CEO, MBAs and Pharmacy District Managers would fight like demons and there would be casualties on both sides. Think of how a fight to get unionized would pan out.

There are people at the top of the pyramid. I think that our goal should be to flatten that pyramid. Effectively putting the brains, talent and experience of the people down below closer to the top.

Our objective should not be to force the executives out because the concentrated power would just move to new hands and, most likely, meaner and tougher hands. The failure of a CEO, for example, makes it necessary for the board of directors to hire a new CEO who could whip the company back into shape. Board members are not pharmacists. Boards can view the pharmacy as just another department that sells a commodity. Chances are that the new CEO would not necessarily be pharmacy-friendly.

I worked for a company that was left almost rudder-less when the son of the founder cashed in and sold out to a much larger company. The problem was that the much larger company had just been through a take-over that left the new expanded company with a whole new executive suite. The fix-it CEO was from the grocery store industry. He treated the pharmacy like an illegitimate step-sister. He neglected us because he didn't understand us and the pharmacy business. The pharmacies did not perform well and, of course, it was the pharmacy managers who took the blame. We pharmacists spoke longingly about the son of the founder who sold out, took the money and ran. The CEO you know is better than the CEO you don't know.

Shortly after the new CEO took over, he brought in a hand-picked executive team. They were primarily from the grocery store industry and they claimed that they were going to turn a pretty good drug store company into a grocery store clone with strong pharmacy influences.

The first month the Pharmacy District Manager showed up. I liked this guy. We got along well. He respected my abilities and I appreciated that he was a very hands off middle manager. He did not look good. There were bags under his eyes. He was exhausted.

"Don't say anything, Jim. Just listen for once." He pulled some papers from his briefcase. He put them on the counter and looked at me.

"Are you going to tell me something bad?"

"Just listen. He managed a weak smile. "Your job as manager is a three-sided pyramid. (I found out later that the three-sided pyramid was right from the CEO. His personal analogy). You have to achieve three results. Inventory, prescription volume and gross profit." He gave me the goal numbers.

"How can I do that? If I keep an inventory that slim, we'll be out of stock all day long and that volume number is way too high. K-Mart just opened their pharmacy and they are less than a mile from here."

"Just shut up and sign here." He indicated a spot on a legal-looking paper. I looked and it was an agreement that I would accomplish the goals in three months."

"That's bullshit, I'm not signing it."

"You have to."

"No, I don't."

"Come on, Plagakis. Do you think I like this? You are a good businessman, but I know that you won't get 24% gross profit. For now, we have to humor the assholes."

I was not about to humor any asshole. I did not sign anything. A good lesson. Unless there is something in it for you, never sign anything. I never heard another word. I spent a lot of time looking over my shoulder though, waiting for a hammer to fall. All of a sudden, I was working in a situation where everyone in the company was merely surviving.

Our aim should be for a change in the framework of how pharmacy is conducted, not a change in the substance of pharmacy. We will continue to do the same things, but our behavior will be looked upon as professional by everyone in the company, including the CEO. That would be revolutionary. That would mean that your use of professional discretion would be respected. That would mean that you would be viewed as a professional and the 15 minute guarantee would not be possible. You can tell people who are running the Prescription Mill how to do it and how long it should take. You cannot tell people who are providing a professional service how they should do it or give them a time limit.

In that context, our revolution will pick up steam when pharmacists stop believing one thing and start believing something else. It will not take time, effort or struggle. The opportunity is ripe. Pharmacists have been aching for this chance. There will be a quantum leap from the old way of thinking to the new. Pharmacists simply need to know that they have permission to transform their own self image. If all they can imagine is seeing themselves as Prescription Mill managers, then so be it. Most of us, however, are very clear that we are something more.

**Our revolution is not something that will be created by an elite group. It will be carried out by ordinary pharmacists, when we change the way we think and work.**

Company policy and procedures are rules are just so much floo flaw. They were not handed down by God and set in stone. They were invented and written down in the dust. If they do not serve the needs of our profession or if they impede the practice of pharmacy, they can be blown away in a stiff breeze. That breeze will be your assertion of individuality as a professional. Every discretionary act in the service of your patients is profoundly revolutionary behavior.

Everything a pharmacist does that asserts individuality and autonomy is in itself behavior that is outside of the company's box. The executives have to know that they are not dealing with gift shop employees. The executives at *the company* certainly don't get it. They do, in fact, treat pharmacists as if they were merely clerks in the camera department. Actually, pharmacists at stores owned by *the company* are treated worse than the people up at camera. Pharmacists don't get rest or meal breaks.

We are living in an historical era of unlimited possibility. The old world of pharmacy is dying and we really do not have a good handle on the new world. It is time for pharmacists to assert themselves and take the lead in creating the innovations that will define the new world of pharmacy. If we don't step up, non-pharmacists will do it and we know what happened the last time we allowed the corporate architects to decide what our job was to look like. We got a running-out-of-control-multi-tasking-stressed-out twelve hour shifts at the Mill.

Most drug store corporations tightly manage and manipulate everything. The ideas that come out of the executive suite at *the company* are touted as the latest and greatest, but are really just the old and moldy schemes dressed up in new clothes. The thinking of those at the top of the pyramid goes largely unquestioned. The 15 minute prescription guarantee was historical because there was a serious challenge. It did not get a free pass from the pharmacists in the stores.

How will the revolution proceed? I can tell you where it will not come from. You won't find it in the alphabet soup. It won't be the APhA, NACDS or NCPA. Heaven help you if you thing that the NABP will start anything.

*It will not come from the universities. The schools of pharmacy are engines for running the status quo. I agree that it shouldn't be that way. That is counterintuitive because academia is supposed to be the breeding ground of innovation and new thinking. The schools* **could** *take the role in creating the space for a quantum leap in the job of working as a pharmacist, but don't count on it.* **The colleges of pharmacy gorge themselves on corporate money.** *The private interests of the schools are more important than the interests of the profession. We are all adults. We know that when you take money, you are beholding to the donor.*

*The schools have laboratories paid for by the drug store chains. Wal-Mart pays for the pharmacy computer systems in some of the schools. Take a tour of the schools and you'll find corporate brands everywhere. Do I expect it to change? Hell no! Does it have to change to allow for a sea change in our jobs? No!*

The transformation will originate with the grassroots. It will be the pharmacists in the stores, where every day, common men and women decide to discover their individuality and seize control of the professional aspects of their job. This is a territory that the rulers cannot control because it is too diverse and too decentralized.

I can't imagine even the executives at *the company*, as idiotic as they can be, thinking that they can tell you or me how to behave as a professional. The Prescription Mill? Yes, of course, that is analogous to the camera department. I will follow the Mill template given me to the letter.

However, when I step away from the Mill to provide professional services, whether it is counseling, immunizations, OTC counseling, or running my fingers through a little girl's hair to assure a traumatized mother that her child does not have lice, I will be independent and self-regulating. My knowledge, experience and expertise are all mine to share as I see fit. When I am acting as a professional, I belong to no corporation.

The possibilities are already embedded in the context of pharmacy. The revolution will not suddenly materialize out of thin air. It is already very much alive. It is manifested by whining and complaining about working conditions. The collective consciousness is ripe for change. We must learn to take ourselves seriously. Our wildest dreams and plans possess real power. The controlling force of *the company* wilts when your counseling contributes to the health and well-being of a patient, perhaps even saving a life. Coercion from the executive suite at *the company* will shrink to insignificance in the face of a single pharmacist with a profound belief in a revolutionary idea.

What the fuck is Fullovit going to do when you spend twenty minutes with a marginally intelligent teenage single mother and she still can't get how to use the Xopenex MDI for her two year old? You refuse to dispense and when the store manager has a shit fit, you don't budge. It is none of his concern. The doctor is ecstatic. He calls the company and tells them that he has finally found a pharmacist who puts the patient first. You end up selling a nebulizer and ipratropium and albuterol ampoules. It is a no-brainer. When you start practicing pharmacy regularly, you will get your profession back in your hands.

**You are at the bottom of the funnel. No real medical care takes place until you dispense the medicine. The law is on your side, actually the law demands that you behave as a professional. How long do you think they will allow you to flaunt the law? You have enormous leverage and power. Use it.**

# Disintegration or Dis-Integration
## The end of Fullovit and his *company*

It wasn't because of George Fullovit that *the company* survived. It was because of the pharmacist middle managers and the pharmacists in the store that the board of directors took *the company* into reorganization. The stock price was less than a 20 ounce Diet Coke. Pharmacists in the stores were quietly disobeying *the company's* policies. *The company's* credit rating had been downgraded by Moody's. The state board of New York outlawed the 15 minute Rx guarantee and every other state board followed within six months. It was like a feeding frenzy. The disintegration had started. Even states where there were no PriceWorld stores jumped on.

*The company* had more than 6,000 drug stores and would have collapsed had Fullovit and his MBA insisted on a fight to the end. The death spiral had started and other predatory industry giants were jockeying to buy the stores in the best locations. The board of directors showed wisdom when they refused a highly leveraged offer from a smallish outfit called Billy Bill's Discount and Pharmacy. The minions of the large chain drug stores knew that the loss of 6,000 stores would damage the industry, perhaps irreparably. The void would be filled by non drug store outlets that ran pharmacies. Big box stores and grocery stores would benefit.

With Fullovit gone, there was an influx of capital and the disintegration of Priceworld became the dis-integration of the company into two discrete entities. The board of directors made the decision that no non-pharmacist would have any authority over pharmacy operations.

That's a very pleasant fantasy. Possible? Of course. Anything is possible when intelligent people are involved. Probable? I am the eternal optimist, but does a company with a stick price of $1.03 per share as I write this have any chance at all? Especially when they keep on doing the same old things, just calling them something new. There will be plenty of drugs to sell and more people clamoring to buy them than ever, but the selling of the product is not the future of pharmacy.

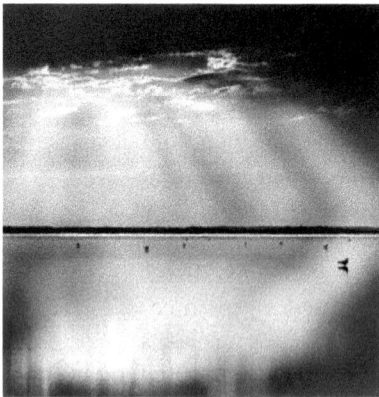

Those of you who know my new age sensibilities and have problems with them, please feel free to skip ahead. Those of you who think miracles are possible and want to have a little fun, strap them on and let's go for a rocket-ride

# Belief

*Be careful what you believe because it is true*

Belief is truth. What we believe is true because we act based on our beliefs and our actions shape our world, especially on the job. If we believe that *the company* has all the control over our days then that is the way it will be. However, you have to understand that we make our own values. We construct the framework of our jobs. Too many of you believe that your job is a piece of shit and that you have no say in how you behave. Some of us know that pharmacy is a profession and we behave as professionals, independent of *the company*.

Belief precedes knowledge. We learned about life in the pharmacy through the filter we started looking through as an intern, working for *the company*. This filter has always been a dark oppressive green color. We firmly believe that a job with no meal breaks is acceptable. We believe that we better keep our head into the Prescription Mill because they are watching. We believe that *the company* will defend us when we get cited by the state board for non compliance with pharmacy law. Yeah, sure! You are all intelligent. It is very easy to agree with facts that support your beliefs. I am presenting new ideas that call your old green-tinted beliefs into question. Allow yourself to see clearly into a new reality. A new fresh vision of what the job of working as a pharmacist can be.

Belief is powerful. What pharmacists believe is the blueprint for the kind of job they have. A culture that believes that pharmacists are glorified clerks will never have the kind of job satisfaction that comes with a profession. A culture that believes that *the company* will orchestrate every single move of a pharmacist will see servitude. If you believe that your job is simply to make money for *the company* your life will be competitive and materialistic. When you finally believe that a patient-centric practice of pharmacy is your goal you will be on your way to professional satisfaction.

Power is belief. There was a point about thirty years ago when pharmacists gave up the social power of a profession and each of us set out on our own. *The company* made this very easy. They paid us well, extended the cloak of a large organization as protection. We gave up our autonomy for security. We allow the power of our profession to be seized by very few people near the top of the pyramid. This is all the more startling when we realize that the power we gave up is based on simple belief. We allowed ourselves to be convinced George Fullovit was stronger, more knowledgeable and more powerful than we were. Indeed, he was just a man and, to make things worse, he was all alone. The fatal flaw was that he was not a pharmacist. We, through belief, allowed our own enslavement.

Belief is irrational. Very few of us, perhaps none of us, examine our beliefs critically. We have accepted what we were taught in school about working as a retail pharmacist. We work with preceptors who believe the same company floo flaw that they pass on to us. We believe that working alone for up to fourteen hours straight with no breaks is the nature of the job. We believe that it is acceptable for us to tolerate abuse from customers who have never been educated that pharmacists are very well trained medical professionals who have standing in the community and deserve deference. We believe that the working conditions define the profession of pharmacy when it is the job that deserves our disdain.

Human beings are irrational by nature. It is not surprising that many of our unexamined beliefs about our job are irrational. But irrational beliefs can be positive and creative if they are examined. When you examine the irrational belief that part of the job is holding it until you wet your pants, you are going to get better. The state of minor mental illness will get better. Irrational beliefs can often be detrimental. An example of a destructive irrational belief is the one that most of us carry around. That we are nothing more than employees of a large corporation. In fact, we are professionals with discretion. When you can see the distinction, your job will be transformed.

The corporation is a social necessity. It is a model for organizing pharmacy when the business has many demands put upon it. There has to be a leader in a hierarchical model. The problem is that the emperor too often has no clothes. An example of a leader who has caused untold misery for thousands of employees is our own George Fullovit. When the very top of the model is so rotten the poison seeps down to all levels. When you are considering a new job, in a new company, you might want to examine the career of the CEO at that company..

`        Belief entails responsibility. When you choose your beliefs, you can't just blindly accept all of the floo flaw about the job of a pharmacist that has been handed down since 1970. If you have control of your beliefs, after examination, those beliefs can shape your professional world. Why not believe in something positive. Instead of embracing drudgery, disrespect and being dominated, why not create beliefs that are positive. Rather than carrying beliefs that grind you down and bleed you of dignity and self-respect, why not choose beliefs that are life affirming?

        Beliefs can change, you know. Not easily, but it can happen. The intellectual arguments that I have presented here may not change anyone's beliefs right away. However, if you now are willing to examine your beliefs about your job, a difference can be made.

# The New Order of Things

*How can our job change if all of the*
*ideas come from George Fullovit?*

We are living in an era of possibilities. The old ways are dying and the new ways are just now beginning to be conceived. There are corporations that are embracing utilizing the pharmacist as a valuable professional resource. M.T.M. and immunizations are just a beginning. It is a process. We are a decade, at least, away from the end game. Our problem is that corporations like *the company* are acting like an anchor. They drag us down. They waste our time and energy. They can't get beyond the past when selling a product was the entire game.

If you are going to contribute to the revolution you have to start thinking like a revolutionary, a rebel of comfort. You will have to look at everything with fresh eyes. When they say that the new idea is the latest and greatest, you must examine the claims critically. You are the expert with your feet on the floor. You are the final arbiter. You can make the MBAs shiver. The dead ideas of the people with the power usually go unquestioned. If the ideas touted as new are just the old and moldy dressed up in new clothes, you have a responsibility to ask questions, at the very least. There is too much that we do in retail pharmacy that goes unquestioned.

The very best CEO is a good listener. He asks questions. The executives of the best run companies know that new ideas originate where critical thinking flourishes: the grassroots. In companies where the executives do all of the "thinking", the grassroots pharmacists are miserable. They hate their jobs. They blame the profession of pharmacy for their misery.

It is merely an illusion that the ruling elites, the executives, control anything. Pharmacists who are controlled by force and the threat of penalties are not motivated to implement the programs that come down from the executive suite. They are much too engaged in surviving a bad situation and, at best, scheming about how they can get out and find a good job. At worst, their scheming is all about how they can quietly undermine and subvert programs that might cost thousands of dollars to be put in place.

Take just a moment to remind yourself of what you have created for yourself. What you are not is a piece worker in a factory. You are smart. Nobody even gets into pharmacy school unless they have the intelligence to get all the way through. You have the ability to think critically and now the permission to examine any and all beliefs about your job. You can make a difference in your own life.

You are a medical professional and the expert on drug therapy. You have an important place and role to play on the delivery of health care. It would be foolish of you to continue to put managing the Prescription Mill ahead of delivering important pharmacy services. Remember, you have legal mandates as well as ethical responsibilities. Living up to your personal standards is where satisfaction lies.

Pharmacists must learn to take themselves seriously. What you believe in is true. Your wildest dreams for your professional life are true. Your most ambitious goals are real and possess real power. The controlling force of all of the corporation's penalties, propaganda and coercion shrinks to insignificance in the face of a single human being with a profound belief in a revolutionary idea.

I could probably continue with talk about the *revolution*. I do not think I want to do that. It would be redundant. I have said everything that I need to say. I am not going to give you a step by step method to professionalism. That is what George Fullovit has been trying to do over at PriceWorld. The idea that you and I have to be told how to act like a professional is an insult.

I trust that most of my readers are pharmacists. That means you are smart enough to get through pharmacy school. I don't think that I need to elaborate any more than I have.

I will leave you with some very simple advice. Believe in yourself as a well-educated medical professional and then go out and act like one.

"How does it feel
To be on your own
With no direction home
Like a complete Unknown
Like a Rolling stone"
Bob Dylan

# The Dangerous Book For Pharmacists

In "The Dangerous Book for Pharmacists" Mister Plagakis looks at the passive nature of pharmacists and how this meekness does not express the real role that pharmacists have earned, but too often do not play.

Pharmacists are well-educated and highly-trained medical professionals, but, in the retail setting, the practicing of pharmacy is regularly relegated to monitoring The Prescription Mill. This is not what will define pharmacists in the 21$^{st}$ Century. Supervising the Prescription Mill will define the Advanced Pharmacy Technician.

Counseling in all of its various forms is the activity that will define us in the 21$^{st}$ Century. Doctors will seek the counsel of some of us. We will provide MTM. There will be pharmacists who prescribe the drugs. If pharmacists want to be useful, they will counsel their retail patients on their daily prescriptions.

Pharmacists will always be the point person for triage for a certain class of patients. They will come to you because you are accessible and because you won't charge them. You are the last medical professional standing who is still free.

Jim believes that it is time for pharmacists to enter the 21$^{st}$ Century and leave the olden days behind.

This book is not, however, about counseling. It is not about getting away from The Prescription Mill, the company's timers and status reports. It is about acting like what you are. A highly-trained medical professional who will make a difference in your patients' lives every single day.

The re-invention of your professional self will not be easy for any of you. There are pharmacists who have spent twenty years running The Prescription Mills for various chain drug store, big box and grocery store companies. They are institutionalized and changing the way they view themselves will not be easy.

It will take only courage for students and young pharmacists to invent themselves in the mold that the 21st Century will demand. They have been provided all of the tools needed. All they have to do is use them. All they have to do is be willing to become a Dangerous Pharmacist.

I work part time for a major chain drug store company in Galveston, Texas. The store is within easy walking distance from The University of Texas Medical Branch. UTMB is a major medical school. I get the opportunity to talk with patients who are students, residents and faculty members. I have noticed something that seems to be lacking in both pharmacy students and pharmacists of all ages. That is Pride.

Medical students seem to be educated, acculturated and indoctrinated with the belief that they are special, superior and frankly better than other medical professionals. I suppose that that was important when doctors were very high up at the tippy top of the medical pyramid, but that kind of positioning is not helpful in the new age of poly-medicine. The pyramid is just about flat. To best serve the patient, collaborative medicine would be the best model. The expert in drugs is the pharmacist.

I don't know many veteran pharmacists who are genuinely proud of their practices of pharmacy. I have not met very many students or young pharmacists who are cocky in the least. They are educated for six years. They have earned the degree of "Doctor". They know more about drugs than any physician will ever know and they still sequester themselves at The Prescription Mill for eight hours a day. They seem to resent it when a patient requests information about prescriptions or OTC medications. They act as if their situation is hopeless. They seem to believe that there is no way out. They have the embarrassing habit of blaming the profession when the only one to blame is the image they see in the mirror in the morning.

There must be a change in viewpoint. There must be a transformation of self-image. All it will take is paying attention and some decent amount of audacity and guts.

Your presentation is important. You are a medical professional. Look like one.

That is not to say that you have to look like anyone other than yourself. You do not have to wear a shirt and tie to look professional. You can wear pants instead of a skirt. You already know what professional appearance is for you. Do that. Whatever it is.

I had the pleasure of working with a young Pharm D in the late 1980s. Cheryl M insisted that her nametag read "Doctor". This made the non-pharmacist store manager crazy. He was a male elitist and Cheryl took him to school. She got her nametag simply by not wavering. She was decisive and resolute. Cheryl got along fine with the manager. He was not a threat. She treated him the way you would treat your car mechanic. He was necessary and she tolerated him. She was the pharmacist and no one was ever to forget it. John D knew that he had met his match and he stayed away from the pharmacy when she was on duty.

There is absolutely nothing wrong with a pharmacist being glamorous if that is the presentation one wants to portray. Cheryl M was the most professional pharmacist I ever worked with, and the most glamorous. She had a streak of henna red in her hair. She wore high heels to work every single day, even the twelve hour shifts. She always wore a skirt and a fashionable blouse. Cheryl M loved what she did and I think it is because she was practicing her profession on her own terms.

I did not follow my own advice during my career. I had a management job where I ran a small drug store for a Los Angeles company. My store was near San Francisco. In the early 1970s, I wore the San Francisco uniform of jeans and tee shirts to work. I wanted to look like a laid back weekend hippie. I rejected the accepted dress standards of the drug store culture. I was into peace and love and grooving. I played rock 'n roll in the pharmacy. I always started the day with Jimi Hendrix "All along the Watchtower" and found space for Janis Joplin and "Piece of My Heart" numerous times during the day.

I don't know if this was a professional presentation or not. It was a very confusing time. There was a local doctor who played the flute for his hypertensive patients. A well known surgeon drove a Harley with four foot handlebars up in the air and a tiny wheel extended way out front. My jeans and tee shirts were probably a mild statement in the social revolution. The patients did not seem to mind. But that was forty years ago.

For me, a professional presentation in the 21st Century usually means a white lab coat over a shirt and a tie, but that doesn't mean that it would work for you. I believe that every single one of you knows what professionalism looks like for you.

There are neighborhoods where a white lab coat can intimidate the very people you want to help. What is appropriate professional attire in Livingston, Montana would be a costume in Beverly Hills. Skin tight may not be the best choice. Cargo pants should be saved for the hiking trail. Just use your head. A professional presentation is critical, but you don't have to sacrifice your personality.

Comport yourself as a pharmacist, act like you are an important part of the process, not as a prescription mill caretaker or as a dude looking to hook up.

The health and welfare of people depends on your making the right call every single day. By law and by tradition, you are the last medical professional in the chain. You are at the bottom of the funnel. What you do makes a difference. The best therapy can fall apart if you do not do your job.

Your first job is to act like you know what you are doing. Being tentative will not engender the confidence or the respect of your patients. You would be wise to act with a certain amount of gravitas when you are with patients. You may not want to have a solemn or serious attitude or way of behaving when you are out on Friday night, but when you are talking to a woman who is genuinely frightened by what her doctor has just told her you will serve her and your image if you act with solemnity.

I know a really competent pharmacist who acted like he believed that he was the second coming of John Belushi. He seemed to think that it was his job to help the patients unwind, to relax them, to get them to not be so serious. He had a joke for everyone and didn't seem to notice that he was often the only one laughing. His personality was terrific for the store Christmas party or his family's Sunday dinner table. His act wore thin every single day at work, but he didn't figure it out until he almost lost his job, fifteen years of seniority and everything he had worked for.

The Pharmacy District Manager showed up one day unexpectedly. He asked the pharmacist to join him in the manager's office. Our pharmacist's face was ashen-white when he returned to the pharmacy an hour later. He held a file. It contained copies of fifty or more letters of complaint. The District Manager had been holding them because the pharmacist was a dependable and loyal employee. Then came the complaint that could not be buried.

Our pharmacist spotted a good looking woman walking to the pharmacy. He went to meet her. He liked her and he thought that his funny man act was something she appreciated. She was a Zyprexa patient.

"Well, Joanie", he said cheerfully, "Who are you today? And who is your handsome escort?" He looked into space. She had no escort. When she was off her meds, she had a special friend ala A Beautiful Mind.

"What did you say?" She had a sly smile, but her lower lip quivered.

"Hardy har, Joanie. Relax. I was just funning you."

She screamed. "Get away from me you perverted creep." She actually tried to hit him. She screamed again and stormed from the store. She never came back, but her letter was written with a poison pen.

Our friend learned a very hard lesson, but it took years and a dramatic event to get it into his thick head that he needed to behave with a professional's comportment.

You are in a position of authority in the pharmacy. You may like the technician's personally. You may enjoy their company as you work. You may go to parties with them. However, you are the pharmacist and they are the technicians. At work, it is imperative that you assert yourself as the last word regarding professional matters. I have seen pharmacists who are intimidated by older veteran technicians. No matter how long you have been in practice, you cannot allow this to happen. I have fired two technicians during my career because they were aggressive middle-aged women who did not like being told what to do by a young pharmacist. My actions were precipitous. I gave no warning. That may or may not have been the best way, but it is what I did. The techs who stayed, by the way, bent over backwards to give me what I thought needed to be done.

Your demeanor is vital. If you are young, it is critical. You may be a Doctor of Pharmacy, but the patient sees a very young person. All it will take is one false move and you will be written off. Once they spot the ring in your lip, trust me, you lose them. Talk in the patois of young America and they will ask for the real pharmacist.

Act like a medical professional. The suit may be too big at first, but you will grow into it and, as you master the role, there will be latitude. You will be able to joke and kid appropriately, but not all the time with every patient. You will learn discernment and that is a very good talent that will serve you well in all areas of your life.

You are what you pretend to be. After a little while, it won't be pretending anymore.

Older pharmacists may be in stuck in a rut. Bad habits perpetuate automatically. If the job has beaten you down, and you act like a Sad Sack, it will take a force of will to change. Confidence is not automatic. It is something we learn and unfortunately preceptors have dropping the ball for decades.

There are preceptors who seem to take pride in taking a student with high ideals and hammering the square peg until it fits in the round hole. They think that there is a certain way of doing things in retail pharmacy and the sooner you get with the program the better off you will be. As far as I am concerned, this is a violation of trust.

The student should get as far away from this preceptor as soon as possible.

I have acted as a preceptor during my career and I am guilty of letting the students down. I work part time at a pharmacy that is within walking distance of The University of Texas Medical Branch. UTMB is a major medical school. I talk with students and residents every day at work. It is apparent that they are being taught how a doctor should act. Even first year students display a certain superior panache. They stand tall, make eye contact and question my choices when I help them make OTC decisions. These kids know absolutely nothing and they will still run their doctor act at me. I like it. They will be flying the medical system airplane even if they will have to share the cockpit in the 21st Century and I want everyone up front to be confident and competent.

Young pharmacists (old pharmacists too) often behave like little children who have not been taught how to act around adults. I blame the pharmacy school faculties for this first. There are exceptions, but most professors don't seem to think that how a pharmacist behaves is important. There is a professor at the University of Georgia who is the exception, but I personally know of no other.

What would be wrong with a P-1 class that expressed these messages?

1. You are beginning a career as a well-educated, highly trained medical professional. Act like one.

2. You will be the last word on drug therapy. Accept that role.

3. You will be the foremost expert on all drugs, that includes Rx-Only and OTC. Act like it.

Why not tell these kids that they need to ACT like they are important professionals and not simply prescription fillers. The retail system is not designed to allow them to easily be important. The schools and the preceptors must give them permission to ACT like they are medical professionals. When they get their first job with a big box store, there will be strong forces playing against them. The Jailers will want to keep them in their place.

I was a preceptor three times in my career. My students were all trained by me to do well in retail. However, as I see it now, I was a dismal failure. I believe that a preceptor should assist the student in aspiring to the highest standards. I didn't do that. I taught them to do what I did. At the time, I was a Prescription Mill caretaker. Most preceptors are equally disappointing.

Students see how their preceptors act and model their own behavior after what we do. They are effectively taught how to ACT as a second class medical professional. We show them that it is normal to have to work a twelve hour shift with no meal or rest periods. We tolerate bad language from abusive customers and the student sees, over and over, that this is the way it is.

How many preceptors have seen the student watching as they allow themselves to be intimidated by a doctor, even when the preceptor is correct?

"I'm.. Uh.. I'm sorry to bother you doctor, but I believe that the dose is.."

The preceptor's face reddens when the doctor shouts at her and hangs up the phone. She looks at the student and shrugs.

The student says, "You're not going to fill it that way, are you?"

"I have to. The doctor won't change the directions."

The patient is put in danger. The student feels helpless and the dangerous slide down to second class medical professional is only iced up and slicker than ever.

Preceptors have a responsibility to strongly express these messages: Do not do it like I do it. Do it like I say to do it. Do not believe that a twelve hour shift with no breaks is normal. This is an aberration that has been perpetuated for more than three decades. The non-pharmacist store manager runs the store, not the pharmacy. I know that I dress frumpy, but you don't have to. You are a doctor. Dress like one. And on and on.

It is immoral to allow new pharmacists to hang there, twisting in the wind, with no instruction on how to act like a professional. All of us veterans have a responsibility to a profession that is in transition. Pharmacy will only be as strong as its practitioners. We need to support all pharmacists in acting proud and competent in any and all circumstances.

By tradition and according to the law in the United States America, the pharmacist must be involved with the delivery of every single drug. Think about this: 90% of all routine doctor office visits end up with one or more prescriptions being written. The pharmacist is at the bottom of the funnel. You are where the rubber meets the road. Without you, everything stops. What about that kind of power is so hard to understand?

However, you have to give up the identity that has colored our profession since the inception of Durham-Humphrey Amendment in 1951. For sixty years, pharmacists have been labeled as specialized clerks who sell drugs. We will continue to sell drugs, but our ability to make a good living should be determined by professional considerations and not simply the selling of drugs. The door is open. You just have to walk through. There will be people who do not like seeing you step up. They will think that it is not your place to act like a doctor. With generosity, I say that that is just tough.

It is all an ACT. Lawyers have an act. Dare to ask them a legal question at the grocery store and you will get a bill in the mail even if the answer took only thirty seconds. Doctors ACT in a certain way and they are given deference and respect whether they deserve it or not. They come into practice already above the line. Unfortunately pharmacists are left on their own to learn how to act. Too often, it is the milquetoast role that fits. That is criminal. It must change.

In the meantime, pharmacists who got no support in this area from their schools or preceptors will have to make it up, do it for themselves. Find your model and do that. Do it with pride, a little arrogance, and some superiority. You are a highly trained medical professional. The expert on drugs. Start acting like it.

There will be bullies

There are people who will try to harass you. There will be abusive people. Bullying is harassment, as are Ageism, Racism and Sexism. If you are discriminated against because of your religious or political beliefs, it is harassment. Be ever watchful. They cannot do this is 21st Century America and get away with it.

Sexual harassment is the signature abuse that gets most of the attention, but there is other, more insidious, harassment that you should be looking out for. I knew a young female pharmacist in San Diego whose career with a drug store company was ruined by insidious and subtle bullying. Her mistake was not reporting the situation to the company's human resources department.

San Diego in the 1970s was a swinging place. This girl arrived at her dream store, close to the beach in the neighborhood where she wanted to live. She was the quintessential young San Diego female. She worked out and ran regularly, was in terrific shape and was Hollywood good looking. The store manager saw her coming and asked her out the first week. She politely declined, saying that she thought it was a bad idea to date people from work. The bullying started almost immediately.

The pharmacist told no one, but the manager expressed his hurt pride to someone in the store whom he trusted to keep quiet. Within a day, employees were making jokes about the manager who had been shot down.

Within a week, it started. The manager told the pharmacist that she had parked in the customer area and demanded that she move her car to the back of the store. She moved, but her car was the only one in the back. There were no night lights. When she came back in the store, the manager angrily accused her of breaking the law by leaving the pharmacy unattended. From that day on, she had to move everyone out and lock the door just to go to the bathroom. And on and on! After six months, she quit her dream job.

A very heavy-set female pharmacist was bullied by a floor manager who liked to make fat jokes. She put up with it stoically until the manager crossed the line one day. The pharmacist was not obese. She had a genetic endowment of big-bones and big muscles to go with them. The floor manager entered to pharmacy and called her a derisive fat name. She grabbed him by the wrists and refused to let go. He struggled and threatened her, but she just smiled. Finally, she told him in no uncertain terms that she would go farther if he ever bullied her again and threw him to the floor. He hit his head on an open drawer, suffered a concussion and had to take two weeks of disability. The pharmacist was suspended for a month without pay.

You never have to put up with bullying. You are a pharmacist. The company considers you the most important employee in the store. You must report bullying to the bully's supervisor. If there is no satisfaction, go to the E.E.O.C.

Any supervisor who commits sexual harassment is an idiot. Modern companies do not tolerate even a dirty joke. Still, there are idiots out there and victims that are even more idiotic. They smile, shake it off and hope that it never happens again. Inappropriate touching is the poster child of sexual harassment. I worked in a store where the manager touched every woman. He was having sex with the bookkeeper and two of the out front clerks. It was as if he had his personal harem. Everyone was afraid to say anything for fear they would lose their jobs. Until one woman was not afraid. She was a new hire, a mature woman who had been around the retail world and knew sexual harassment when she saw it. She called Human Resources. Within days, company security escorted the manager from the store.

No matter what the harassment is: Racy language, jokes, touching, suggestive whispers. You must tell the harasser that the behavior is unwanted. After that, if the harassment continues you have a legal case that will be pursued vigorously by most companies and by the E.E.O.C. if the company drops the ball. You would be wise to express that the harassment is unwanted in writing directly to the harasser, with a copy to his/her boss. Do not let embarrassment stop you. If you do not take action, you will be a victim forever.

Bullying can be subtle and pervasive and difficult to report on. A small thing such as criticizing your for going to the bathroom too often is bullying. How do you report that? You have to stand up for yourself. You are probably more intelligent than the bully. Be smart about it.

My perfect job was in Oak Harbor, Washington. My house was less than ten minutes from the store. I was the Pharmacy Manager and wrote the schedule. I had enough seniority that I could get my vacations in the summer. I was well respected in the community. What about that job could make me quit?

The store manager did not like me. He wanted the quintessential lick, stick and collect-the money-pharmacist. My patient centric practice rubbed him the wrong way. I spent too much time with the customers. I left the store to detail doctors about our pharmacy. He didn't like that. At Valentine's Day, I took big baskets of fun-size candy bars to the nurses. It was a waste of money, he said and I did it without getting his permission. I opened an account with a local nursing home and had eighty new patients over night. He liked that, but did not want me to get over-time for eight hours a month to inspect the home.

But, what he hated the most was that I made more money than he did. He took it personally and tried his best to punish me by bullying. I did not allow this man to manage me. I have always felt that pharmacists should be managed by pharmacists. These were high-testosterone days of my forties. I would not do the same things today. In-your-face may give you momentary satisfaction, but look at the long road.

He wrote me up at least once a month for everything and anything he could find. A customer threw an "F" bomb at me. I walked away and I was written up for being rude. He came into the pharmacy and found some dust and wrote me up for having a filthy pharmacy. I never signed the write-ups, but they work on you. I was the one to finally blink. I brought the boil to a head. At a meeting with my Pharmacy District Manager and the District Manager, I quit my dream job because my PDM was not backing me aggressively.

I was an idiot. The commute for my next job was ninety minutes one way. Be smart. Keep a low profile and follow what I have to say here:

# Document Everything

Document, document, document. Two out of ten of you will do this. Eight of you will wish you had. You are not going to get very far when you whine to your District Manager that your store manager gives you a hard time about going to the bathroom. It is a different story if you keep a diary (written or a document in your computer) where you have recorded multiple incidents of bullying. You will get attention if you have forty-five incidents with date, time, names and what was said.

*10/21/2010 at 2:30PM, the store manager, Dave Johnson, stopped me when I was coming out of the bathroom. He said, "What takes you so long, Jill? You have customers waiting. If it takes you that long, maybe you should see a doctor or get a good laxative." He followed me to the pharmacy and laughed all the way.*

*11/15/2010 at 1:15 PM, the store assistant manager, Denise Biggs, told me to put my sandwich down and get back to work. "You are taking too long to eat and never bring tuna again. It stinks."*

*9/1/2010 at 8:45 AM, the store manager, Dave Johnson, met me at the front door and said, "Put on some frikkin' makeup. You look like shit, as usual." When I started to cry, he laughed.*

Bullying becomes bullying when it is meant to harm you emotionally and when it is repeated. You must document every incident.

Sexual harassment is the golden gift that any person with authority can give you, if they are that stupid. The company could be liable for monetary damages. They could be ordered to pay you for a long time.

A young woman who worked for the FOX Network super star Bill O'Reilly got a reported settlement of one million dollars because she was smart enough to tape record O'Reilly's lewd, lascivious and suggestive telephone calls.

Document sexual harassment in detail. You must tell the harasser that their behavior is unwanted. You must do this in writing. Do not trust a witness. The person with authority has authority over them also.

Running his fingers through your hair is sexual harassment. If she pats you on the bottom and squeezes your buttocks, she is engaging in sexual harassment. Touching of any kind, if unwanted, is harassment.

In the mid 1960s, I worked for a widow who had inherited her husband's pharmacy. One night at closing, when the store was dark, she pushed against me and kissed my mouth. She made it very clear that she wanted sex and she wanted it right then. I was in a marriage that was to get miserable any day, but I didn't know that yet. I declined her invitation and subsequent overt and covert come-ons, but I liked the attention. I was like any boy out there. It felt good to have a woman after me. I laughed when I told her that I wasn't interested. I winked and touched her cheek. We worked together all day long, six days a week. I didn't want to have tension between us and I am not the type of man to be mean, but I loved every minute.

In 1966, she could have hit on me twice a day, every day until I relented. I was a 27 year old boy. She was an attractive woman in her forties. I would not have been able to hold out once my neighbor hinted that my wife was having an afternoon male visitor when I was at work. I asked my boss to stop. What she did was moderate the frequency of her suggestions. Had it been the year 2011, with the laws about sexual harassment, I could have owned her store.

During the early 1970s, I was a pharmacy manager in the San Francisco Bay Area for a small chain of pocket size drug stores that was based in Los Angeles. In the late 1970s, I was also a pharmacy manager for a big box chain store in San Diego that was owned by a large grocery chain. I had ongoing sexual relationships with two technicians during the 70s.

The young woman in San Francisco was very needy. She liked being in a position with the boss where she was protected. We had sex at the store, after closing or before opening. We did it at my apartment and at hers. This girl had a great case for sexual harassment if she ever got tired of having sex with me. At any time, she could have said that my advances were unwanted. That would have been a game-changer. There were no clear-cut laws about sexual harassment in the 1970s

The woman in San Diego was a smart rat. We talked about what a bad idea a sexual relationship between us would be. That was a sophisticated form of foreplay. Once we got together all alone, we acted like two alley cats. We did it everywhere we could and we did it often. Then, one afternoon, after, we were in bed. She had her head in the crook of my left arm. I was using my right hand for the after cigarette.

I remember gagging when she asked me for a raise.

She rolled her head and bit my chest playfully. "Jimmy," she said, "I think it is time for me to get a raise." The words I can recall in an instant they are so indelibly inked onto my brain.

I stuttered. "Lynn, I can't do that. Your review was just two months ago. I gave you a high score. You got a good raise."

She sat up halfway. "You aren't relaxed, Jimmy. Why did you get so tense so quickly?" She kissed my chest and was out of bed in a flash, getting dressed silently. The moving picture of her reaching behind to clasp her bra with two hands is also in my brain forever. She was drop dead gorgeous and I had to fire her. Not right then, but after a day or two of realization of the difficult situation I was in.

Lynn sent a graphic letter to my boss in Los Angeles. He made a special trip down to talk. After he soundly reprimanded me for using such bad judgment, he leered, "How did you get so lucky, man. I'd love a chance with her."

That is the difference between now and then. With proper documentation and the will to get me, my ass could have been grass if those two stories had taken part in the 21st Century.

Most large companies with pharmacies take sexual harassment very seriously. They know that the federal government will prosecute them vigorously.

*Sexual harassment is intimidation, bullying or coercion of a sexual nature, or the unwelcome or inappropriate promise of rewards in exchange for sexual favors. In some contexts or circumstances, sexual harassment may be illegal. It includes a range of behavior from seemingly mild transgressions and annoyances to actual sexual abuse or sexual assault. Sexual harassment is a form of illegal employment discrimination.*

The United States Equal Employment Opportunity Commission considers sexual harassment to be a serious crime. Hearsay and "He said, She said" will not be enough. Complete documentation will be necessary. You can bet that the store manager is documenting every false step you make.

*See complete details regarding sexual harassment, employment harassment and the E.E.O.C. in the appendix.*

# Ethical Practices

Define as: being in accordance with the rules or standards for right conduct or practice, esp. the standards of a profession:

*It is considered ethical for pharmacists to respect the values and abilities of colleagues. A pharmacist places the concern for the well-being of the patient at the center of professional practice. And on and on!*

We certainly have a lot of work to do, don't we? And those are only two of an entire list of principles that we are supposed to follow. We certainly cannot justify twenty-five dollar gift cards when the patient transfers a prescription. The days of cut throat pharmacy are still lingering and the effects are still running the job of working in a pharmacy. We refer to each other as competitors rather than colleagues. We put so much attention on running the Prescription Mill that patients leave the store with prescriptions that could be death-dealing without proper counseling.

Ethics are hard and fast rules of the profession and they must be followed if the pharmacist is to display integrity. Unfortunately, modern day pharmacists disregard the ethics of our profession every single day.

When was the last time you observed a pharmacist consistently *seeking justice in the distribution of health resources.* I am not sure that I even know what that means. I do know that the chain drug store pharmacist distributes every single drug (health resource?) that they can at the highest price they can charge. There is no such thing as *pro bono* in modern day pharmacy. There was forty years ago. We gave drugs either free or at a reduced price to people who could not afford their prescriptions. Does our profession even have a heart anymore?

There is an ethical mandate that pharmacists intercede on the behalf of the patient in the use of drugs. How many of you effectively ignore requests to assist with over the counter products? The Prescription Mill demands so much attention that you hardly look up when you say, "On the left side of aisle nine" when an elderly woman asks where she can find medicine for arthritis. This is not *serving individual, community or societal needs.*

It is interesting to me that the most often quoted Pharmacist Code of Ethics was adopted by the American Pharmacists Association in 1994. The organization with the name that indicates that it is in business to protect the interests of pharmacists does not protect anyone. The expressed purpose of APhA is: *Improving medication use. Advancing patient care.* This is very high sounding, but these six words are worthless propaganda. The working conditions in the job of modern day pharmacy guarantee that patients will use medications ignorant of the dangers and that patient care will be ignored. The APhA has not addressed the pervasive demands that the Prescription Mill puts on the pharmacist.

APhA stubbornly refuses to even make a statement about working conditions in the job of working as a pharmacist even though pharmacists are refusing to renew their memberships. It will take a sea change before the APhA starts looking out for the welfare of its members. *Improving medication use and Advancing patient care* is just so much hoo haw when pharmacists all over the country work 12, 13, 14 hours straight with no rest periods.

How can a pharmacist *advance patient care* when she is exhausted and when her gobbled dinner meal was a giant Snickers bar, Diet Coke and a bag of salty chips? There is no code of ethics that can protect the patient when the pharmacist is exhausted and has wet underpants from holding it too long. The tired pharmacist is a dangerous pharmacist. Our considerable skills are degraded after the eight hour mark with no rest periods.

The APhA must recognize that the current chain store model is not going away. The stated mission of APhA is impossible unless/until the Prescription Mill is modified to allow pharmacists to actually give attention to the needs of the patient.

# Personal Professional Standards

This is where the "rubber meets the road" for all of us. We did not make up the Code of Ethics that was adopted by the APhA. We really cannot be concerned when it is this large, seemingly professionally amorphous, organization that expresses that we are supposed to *promote the good of every patient in a caring, compassionate, and confidential manner*. Who has the time? What are they thinking? Haven't they been in a pharmacy lately?

Our professional standards, however, are 180 degrees apart from Ethics. Our standards are all our own. Nobody else makes them up for us and they are nobody else's business. If we live up to our personal standards, we will feel good on our way home at the end of the day. We will tell our spouse about the middle-age man with the horrible stammer who took forever asking about cough medicine. You did not interrupt him. You were patient. You lived up to your personal standard of allowing all of your patients to have their dignity.

If your personal professional standards do not include patient care, then, what the hell, stay at the Prescription Mill, go home and happily have a few beers while you watch ultimate cage fighting on cable TV.

If you do include patient care as your primary responsibility, what do you think on your drive home when you recall the young Hispanic mother staring at the Xopenex HFA MDI as if it was a ray gun from a science fiction movie? You did not go over to help her and make the excuse to yourself that you cannot speak Spanish. You feel awful as you pass Midway Boulevard because you know that there are always Spanish-speaking employees in the store who could have helped. The picture in your mind of the small child with a wheeze holding on to her mother's leg won't go away. The image is tormenting you and you promise yourself that you will do better next time. But, will you?

The Prescription Mill timers and computer-generated reports tell the pharmacy district manager how fast you fill prescriptions. You can't allow the wait times to increase. Your company is the one with three letters that seems to believe that they can get more productivity from intimidation. You compartmentalize your professional standards and store them safely for "one day". You can do this, but not with impunity. Like a nagging mother, your personal standards will always be there to get you if you ignore them.

My personal professional standards include taking care of poor people who ask me for help. I will do everything I can to service their needs. Pharmacists are the most accessible health care provider and poor people know it. They also know that we provide help and expertise for free. That makes us the primary medical care provider for thousands (if not millions) of poor people who have no insurance and who do not qualify for social programs like Medicaid.

Poor people will find twenty dollars (taken from grocery money?) and they will come to you in hopes that you can sell them an OTC product that will help. You cannot do this in every case. They may really need to see a doctor. You know that they will end up at the Emergency Room because it is free to them (but is a bill of a few hundred dollars to our medical system).

Then, there are the people who are wounded or seriously hurt or very ill and do need to get to the ER immediately. A guy who had been in a fight had two of the other guy's teeth stuck in his knuckles. He asked me to help him get them out. I sent him to the ER.

Usually, however, all they want and need is temporary symptomatic relief while they get over their cough or cold or aches and pains. Toothaches are difficult, but you can give some help. Earaches or diarrhea, they better see a doctor. I am compelled to help these people. It is what I do. If I neglected them, I would pay an emotional price. I cannot ignore my standards. I am the only one who is always there to see me when I neglect my own standards.

I do everything I can do counsel on every new prescription. Rarely do I spend more than thirty seconds on a prescription because if I went beyond that time on the clock, the information begins to get arcane, the patient's eyes roll and they go away. If I neglect to counsel up to my standards, I know it. I can forgive myself a few times, but when I miss the chance to counsel a patient who has never used an inhaler before, I do not feel proud of myself for running the Prescription Mill when all I had to do was remind the patient to rinse her mouth out after each use.  When she gets oral thrush, is it my fault?

Your professional standards are set in stone.  You can try to modify them or deny them, but few of you will be successful.  The longer you deny your standards of practice, the more miserable you will be.  Personal standards are rules that you made up.  No organization has imposed them on you.

*How many roads must a man walk down*

*Before you call him a man?*

*How many times must a man look up*

*Before he can see the sky?*

*How many years can a mountain exist*

*Before it's washed to the sea?*

**How many times can a man turn his head**

**Pretending he just doesn't see?**

*The answer, my friend, is blowin' in the wind,*

*The answer is blowin' in the wind.*

*Bob Dylan*

Your Ace in the Hole

It is astounding to me that the majority of a quarter of a million pharmacists violate the law repeatedly every hour of every single day. They don't just disobey the law of the land, they flaunt their behavior where everyone can see it. You know and I know that we are mandated by law to counsel our patients on their prescriptions, but we still consistently do not comply with the law.

Many of us do not care about the law. Counseling would just be an intrusion in an otherwise uneventful day at work. All we want to do is make sure that the prescriptions are filled properly, eat our lunch if we are lucky and listen to the cash register sing. We are piece workers, just like in a factory making widgets. We do the same thing all day long. The only thing we get out of it is to be able to look at a bloated pay envelope every other Friday.

Your personal standards are the easiest to violate. No one knows that you believe that you should counsel. No one other than you knows that you think that the profession has cheated you. It is not pharmacy. It is the job. You do what everyone else does. Counseling on new prescriptions could get you unwanted attention. If you work for the company that times everything, you could look bad. Pathetic.

The companies we work for have expectations. They have every right to count on it that we will do what is necessary for them to make a profit. However, we are educated professionals. We are not piece workers. As professionals, we are directed to make judgments that can make a difference in the lives of our patients.

I know that many of you are so beaten down that you cannot even imagine that the choices you make when you talk to an elderly woman with congestive heart failure about her drugs could help keep her alive at best and, at least, enhance the quality of her life. You are a Doctor of Pharmacy, for goodness sake. You have the responsibility to *discriminate and use discretion.* That is something that a piece worker, a non-professional is not allowed to do.

There was a time not that long ago when pharmacy was defined by dispensing. We filled the prescriptions exactly the way they were written. We were not allowed to differentiate about anything. We weren't even allowed to put the name of the drug on the label. Counseling was interfering as far as doctors were concerned. This is the 21st Century. We are expected to advise the patient that one thing is better than another.

When you advise your patient that you refuse to fill the prescription for Vicodin because he has just told you that he has Hepatitis C, you are behaving as a professional behaves. You cannot hide behind the doctor.

Just because he is an idiot doesn't relieve you of responsibility if there is damage done. If all you want to do is run the Prescription Mill, you should have been a technician. Trust me, it will not be long and advanced technicians *will* be running all of the routine tasks that the Prescription Mill requires.

I suppose that pharmacists are going to continue to break the law until a few of us get cited and suspended for our blatant disregard of the law. I expect that to happen sooner than later.

The state boards know what is happening. Yet, they look the other way. The purpose of the boards is to protect the public. The law says that we must counsel. How long can the boards ignore what is happening?

The cornerstone of pharmacy practice is the pharmacist's duty to warn. If someone would happen to die because the pharmacist did not sufficiently warn about the dangers of their drug therapy, I can see a good attorney coming after the pharmacist, the company and the state board. The pharmacist has a professional duty to warn. The company knows this and should provide all of the tools, including the time, to see that this happens. The state board must regulate the profession in a manner that protects the public. This would include vigorously enforcing all counseling laws. Who knows when this will happen? In the meantime, cover your own ass.

Employers will always say that the pharmacists are expected to obey both the letter and the spirit of every single pharmacy law. But, with a wink and a nod, the employers really know that they expect the pharmacists to run the Prescription Mill at breakneck speed, just as fast as they can make it go.

A few years ago, I corresponded regularly with a young pharmacist who was determined to practice pharmacy the way she was taught in school. She was a modern day pharmacist version of Don Quixote. This young lady worked for a medium-size grocery store chain. All of her supervisors were non-pharmacists. They knew only that she was rocking the boat with all of her in-your-face professionalism.

They had never seen a pharmacist who had so many demands. The biggest being her insistence on counseling. Our pharmacist was foolish. She purposely overdid her professional tasks. She did more than counsel, she would spend so long that patients would bail out right in the middle. They would walk away in a dither.

"I will call you if I have questions." Smile. "I gotta go now."

What they wanted to say was: "I am so confused my head is spinning."

Our young friend counseled too thoroughly and her patients often tuned out and never heard a word. The company declined to give her a permanent location as a staff pharmacist. They kept her on as a floater, but her assignments were at stores farther and farther away from her home.

There is evidence that one huge drug store company may be ready to break the mold. They say all of the right things. They mean for their pharmacists to be professional partners in the care of their patients. In Texas, where it is required that the pharmacist documents every single act of counseling, this company has assigned a bar code to each pharmacist. When pharmacists counsel, they must scan their own bar code to verify that counseling has taken place.

I have seen really good pharmacists leave their badge with the bar code beside the register. The pharmacist never leaves the Prescription Mill and the technician scans the bar code even though no counseling has taken place. This is legally fraud. The technician is engaging in fraud with the pharmacist being complicit in this fraud.

The company would probably be held harmless in this case. They will say that you have not complied with policy.

*"Your honor, this technician and pharmacist know what our company policy is. They are expected to fully comply with the requirement that the pharmacist counsels and that this counseling be recorded."*

Putting your license on the line and being left swinging in the wind if there are citations handed out. Your company will not defend you if you are caught breaking the law.

The law is your trump card. What is so hard about this to understand? Federal and state pharmacy laws trump anything and everything. They trump your personal standards. The pharmacy laws negate ethics and your company's policies. There is no argument. Your company's timers are like whispers in the wind when the law is involved.

I have labored writing this section. It is very difficult to keep your eye on the prize when there is so much company noise all around you. The timers, inventory, shoppers' cards and sale items all seem to be more important than pharmaceutical care, whatever that is.

I know only this: If you choose to counsel up to your personal standards, the law supports you. If anyone with authority suggests that you counsel too much or that you should not counsel at all, you will have a very good, compelling case to take to the state board. Be smart. Document everything.

# The Ideals and Reality of the Prescription Mill

There will always be a Prescription Mill in retail. We will always dispense a product and there will always be a profit motive. By law and tradition, a pharmacist must be involved in the dispensing of every single prescription.

That is the way it has been since Durham-Humphrey and that is the way it must remain. However, the sheer number of prescriptions our industry will be filling will require that community pharmacies mimic what is done at mail order establishments. That means the use of robots and authorizing specially trained technicians to do much of the verifying that is now done by pharmacists. This will allow the pharmacist to do the tasks that will define our industry in the 21$^{st}$ Century. Counseling, of course, will be our signature service to the patient. You can include MTM, compounding and whatever else the APhA can dream up and they will dream stuff up.

Immunizations will be a significant factor in the practice of pharmacy once doctors realize that the time that shots take, plus the storage requirements and the financial loss of having to discard vaccines that go out of date make immunizations a losing proposition.

Of course, compounding is the quintessential skill of the pharmacist. No other medical professional has been trained to compound. I am reminded of a transaction a pharmacist had with a patient in California. The pharmacist had just told her that a certain compounded prescription would be $80.00.

"But my doctor said that it should be only twenty dollars."

The pharmacist smiled and handed the prescription back to the patient. "Have your doctor mix it then."

"I don't think doctors know how," she complained.

"That's right," the pharmacist said.

Pharmacists do not do surgery and we do not tell the patient what the doctor should charge for his or her services.

There will be more money for you, eventually, from doing specialized tasks. When you show that your efforts are producing a satisfying return on investment, you will be rewarded. There may come a time when you have to refuse. Doing something for nothing is stupid. However, these companies pay you well. You can afford to bide your time.

There must be a shift in paradigm to achieve the ideal Prescription Mill arrangement. Right now, we are employees of the companies we work for. In the future, can you see bringing your practice of pharmacy to an employer for a price? You practice pharmacy your way and you are paid to do it in their location. The distinction is fine, but I believe that this kind of arrangement would be more satisfactory for the pharmacist.

We are not living in an ideal world and it will never be perfect. We are engaged in a profit-making business. The sooner we show the employers that we can partner with them in creating a more satisfying return on investment, the sooner that we will get a piece of the action.

Right now, a premium is put on filling prescriptions and filling them faster than the competition. This has caused a downward spiral that will be difficult to fight our way out of.

# The State Boards

The state boards are not mandated to protect pharmacists. They are not going to do a thing about your working conditions unless you show that those conditions endanger the public.

The state boards of pharmacy are regulatory agencies that are in existence for one reason. That is to protect the public. They do this by making sure that pharmacists are competent with the licensing process. This is a complicated society and the rules the board enforces can be knotty and difficult at times, but be very clear about this: The boards are created to protect the public. Period! Your working conditions are not an issue unless the conditions cause the patient to be endangered.

How can the state boards justify looking the other way when pharmacists work up to 14 hours straight without even a fifteen minute uninterrupted meal/rest break? At about hour nine, the pharmacist is exhausted and very dangerous. Only an idiot would argue that pharmacists are at the top of their games when they are tired. Fourteen hours straight can and does cause harm to the patient.

A fifty-something pharmacist in Washington was a diabetic. He had all of the problems with working conditions that you and I have, but, unlike you and me, this man could not get away with eating junk. But, he tried it anyway. A candy bar and a salty snack were so easy compared to heating up a soup or bringing a salad. He worked one twelve hour shift every week.

One night, he dispensed piroxicam when the prescription called for nifedipine. That only way that is possible is when a licensed pharmacist is seriously compromised.

Does that mean that the state boards deny pharmacist licenses to diabetics? How about regulating the work place so that diabetic pharmacists don't kill people! This man was an experienced and capable career pharmacist, but, on his twelve hour days, he often acted like he had been drinking.

A pharmacist who works that many hours straight expends a lot of energy and needs good food rather than a Big Grab of Cheetos, two cans of Red Bull and a whole bag of Fun Size Baby Ruth bars.

I can easily speculate that tired pharmacists make errors that can and do kill people. Think of it. Theoretically, a quarter of a million pharmacists are engaged in the dispensing of three million prescriptions a day in the United States.

The state boards are like eagle-eyes when it comes to pharmacists drinking while working or under the influence of Rx-Only or street drugs. How can the boards justify not paying attention to working conditions and pharmacist scheduling that makes pharmacists impaired and dangerous? The pharmacist who comes to work at 8:00 AM is not the same practitioner at 9:00 PM. The tired pharmacists, who have been holding it until their underpants are wet, are much more liable to make a mistake.

Let's call in the "Bathroom Proposition". Pharmacists who cannot find the right moment to leave the pharmacy to go to the bathroom are proportionally more likely to make a prescription error than pharmacists who feel comfortable going to the bathroom at their convenience. In simpler words: If you have to hold it, you are unsafe.

A tired pharmacist who has spent ten hours multi-tasking, ordering, managing a runaway Prescription Mill, ringing the cash register, manning a turn at the drive-through and still trying to counsel is an unsafe pharmacist.

Color all of these pharmacists red. The color of Stop. They need to eat and they need to rest. They need to take bathroom breaks. Children need for their pharmacist parent to at least call them on the cell phone on the night of the big game.

Let's speculate that only half of all pharmacists work in race-track pharmacies and end up working hungry and exhausted. That is 125,000 pharmacists who are liable to make life-threatening mistake every single minute of the day.

What about pharmacists who are impaired by working conditions can't the state boards see? They know what is going on in retail pharmacies. They are not idiots. The public is in danger and they are frozen. But why?

Are the state boards of pharmacy in collusion with the retailers? Do they need proof that people die from mistakes like the piroxicam/nifedipine error? Or are the boards simply keeping their heads in the sand so they don't have to see a huge problem? Perhaps they know that if they see that working conditions endanger the public they will have to do something to regulate the JOB of working in a pharmacy to protect the public from pharmacists who are impaired by the job.

The state boards are regulatory agencies. Like most regulatory agencies, the people in the seats are pharmacists with a sprinkling of public members. There are states (Vermont) where the majority of board members are working pharmacists. Look at most states, however, and you will find that most of the board members are pharmacists who are mid to upper level management in the chains.

You can trust that most board members have a nice lunch every day. You can bet that they sit down while they work. They go to the bathroom when they have to and nobody will criticize them for taking so long. They are as far from you as they can get and still be pharmacists.

One of the major chains schedules the pharmacists for 12 to 14 hour days, three or four days a week. One pharmacist all day long. No overlap, no meal break, no bathroom breaks. The state boards know this and look the other way.

The best strategy to get working conditions transformed is to get the state boards to publically acknowledge what they already know. We work in a minefield.

What can you do? Document incidents where you were a danger to the public due to working conditions. When you and your colleagues from around the state have a pile of bad stories, share them with the state board. Make demands. It's worth a try.

# Change is the Air We Will Breathe

# In the 21$^{st}$ Century

The major drug store companies tend to be conservative and change is perceived as dangerous. Let's use the big three drug store companies for our illustration. We will not talk about the big box stores or the grocery stores. They have embraced a kind of change that has not been productive at best and has been an embarrassment professionally at worst. Other than refusing to work for them, there is nothing we can do about it.

The big three drug store companies have been relatively successful putting all of their eggs in the Prescription Mill basket. They have used dollars-off coupons for transferred prescriptions to attract new customers. What they have gotten are customers who transfer back and forth. There is no loyalty. Wait time has become the end all focus. There is nothing more important than having the Prescription Mill run efficiently at lightning speed.

Rite-Aid has instituted a new program guaranteeing a wait time of fifteen minutes or less. This is an example of a conservative upper management that is suffocating. They are not breathing the air they will have to if the company is to survive. One share of Rite-Aid stock sells for less than a 16 ounce bottle of Coca-Cola at the front of the store.

Rite-Aid is just trying to do the same old thing better. It will not work. If they try to do the same old thing differently, it will still be the same old thing. The same goes if the try to do it harder.

There must be a paradigm shift if the drug store companies want to continue to be the first choice for people who have prescriptions to fill. It is my contention that individual pharmacists can be the trim tabs that can cause transformation.

The days when the drug store companies can provide satisfying returns on investment by selling a product are ending and getting this product into the hands of the patient quickly will no longer be the primary criteria for superior service. People in the 21st Century will be informed consumers and they are going to demand everything that we can give them. Unfortunately, they are going to want it for free, but that's a story for another time.

This is the beginning of the sea change that will take over our industry and determine whether a retail pharmacy company succeeds or continue to see their stock sold for the price of a soft drink. At this point, the companies are very big and getting a huge ship to change course is not an easy job. The biggest problem is that the executives tend to see change as a competitive disadvantage. To put more importance on professional services looks like risk-taking. They tend to want to do the same old things, better and differently and, like Rite-Aid, they want their pharmacists to do the same old things harder. Taking loyal, creative, well-educated professional employees and giving them tasks that make them feel like piece-work factory workers will not produce the kind of results that make the share-holders giddy.

My prediction is that the company that encourages pharmacists to practice the profession at a level that satisfies each individual's person standards as well as professional ethics will break away from the pack and rule during the 21st Century.

I practice pharmacy ethically. At least, I think I do. My personal standards are satisfied. I counsel on almost all new prescriptions. I give personal attention and counsel patients on the use of OTC medications. I have been doing this for a long time and have become very good at it. The result is that I have a significant personal following. They depend on me for advice. They want to be counseled.

Yesterday, I counseled a woman on Esgic. I told her that butalbital is a short-acting barbiturate and that the drug contained caffeine and APAP. She asked what that meant.

"You know what a *downer* is?" She is in her early 60s, from the day. Of course, she knew what a *downer* is..

"Is this a *downer?*" Her eyes got big.

"Short-acting," I said, "It is terrific for tension headaches, but it is very easy to get habituated to it. Be careful. The doctor gave you a bunch of refills. Be judicious with this drug."

She made eye contact. "Is this something new? I have never had a pharmacist talk to me about my prescriptions before."

"It's not new," I said, "I will always counsel you. It is something that you should expect."

"Thank you." She said it as if I had given her a whole handful of twenty-five dollar gift cards. "This is my drug store from now on." She looked at me. "And I live on the other side of town."

Because of my thirty seconds of counseling, the drug store gained a loyal patient. I know that it doesn't look like it can be that simple, but it really is. Our only job is to show the drug store companies that our efforts in providing professional services will make them more money than their obsessive compulsive focus on the Prescription Mill and wait times.

The secret will be if they can listen. One company has indicated that its focus will be on pharmacists actually practicing pharmacy. Right now, the programs are rigid. They seem to believe that they have to tell you how to do it. But, that is like painting by the numbers. You are much more creative. You do not need to be told how to practice pharmacy.

Recently, Larry Page, one of the founders of Google, took over the job of CEO. The stated reason is that the company has gotten so big that it has lost its edge. Facebook and Twitter are nimble and quick and Google has become fat and slow.

The big three drug store companies are fat and slow. Rite-Aid is on a downward death spiral. Rite-Aid pharmacists are probably in the best spot of all. Only out-of-the-box innovation can save them from reorganization. The industry could not afford to lose 5,000 pharmacies so no matter how bad it gets, your job as a Rite-Aid pharmacist will be safe.

CVS is in trouble and anyone who has been watching knows it. As I write this, the Federal Trade Commission is investigating CVS-Caremark for violations of anti-trust laws. If any CVS executive lies to the Feds, he/she will end up in jail. Martin Grass, the ex-CEO of Rite-Aid lied to the Feds and ended up in federal prison for ten years. The Feds are bull dogs. It is well know that the federal government does not even come after you unless they already have an air-tight case. If you want to read about a company that cheats at every level, Google "CVS in trouble".

Rite-Aid and CVS are not positioned to even sniff at the air of the 21st Century. Walgreens is. WAG is almost obsessive about following the rules, all rules. This is a huge company and exacting change is never easy, but there is room here for pharmacists to actually practice their profession.

Change, by the way, never comes easy when it comes from the top. The most profound change for the better will always come from the bottom. The companies that encourage intrepreneurship will thrive in the 21st Century. Nimble, light and quick will prevail. Heavy, plodding and slow will fail dismally.

In the end, the company that utilizes its professional staff as more than mere Prescription Mill tenders will win the day. The returns on investment will be eye-popping. The best, most competent pharmacists will want to work for that company.

The game is on. It will be fun to watch. You have amazing opportunities to practice just the way you want to and, trust me, when you show that you are making money for the company, you will get some of it. You have every right to expect some of it, but remember that it is their store. You will have to make a deal. If they won't, it's a new day. Start looking around.

# Competition

As long as we consider each other to be competitors and not colleagues we are going to be stuck in the *pharmacists-sell-a-product* model. This may come as a surprise, but the idea that pharmacists from a different company are our enemies is a relatively new idea. It began in the late 1970s when drug store companies began to undercut each other on price. That was when 85% of all prescriptions were for cash. We need to examine this idea and get over it. If our companies insist on playing the competition game, so be it, but we can still treat each other with deference and respect.

The next time another pharmacist calls for a transfer, stop what you are doing and take the call. They aren't trying to steal your youngest child, for goodness sake. The patient has a right to take their prescription anywhere they want, within the law. The other pharmacist probably considers this task of calling for a transfer a pain in the butt. Just get it over with. Treat each other with courtesy and friendliness.

Treating each other as colleagues is the beginning of creating a professional community that can do amazing things. Think about it. Who said, "We will hang together or certainly we shall hang separately."

## Dignity, Self-Respect & Integrity

There is a very small pharmacy professional organization called *The Pharmacy Alliance*. I am plugging them only because *TPA* has as a stated mission: *A goal of creating conditions that promote Dignity, Self-Respect and Integrity in the job of working in a pharmacy*. I am one of the founders of *TPA* and it is astounding to me that thousands of pharmacists can loudly and vociferously complain about the working conditions at their job and not jump to write a check for $60.00 to join *The Pharmacy*

*Alliance.* At one point, the officers of *TPA* wanted to give up. I would not allow that, however, at least not yet. Time is on our side. *TPA* is still alive, if on life support. You might want to visit www.thepharmacyalliance.com and take a look.

# Dignity

Pharmacy is a profession. You are a highly educated medical professional and it is pathetic that I have to talk to you about the lack of dignity in your job. Dignity is simply the condition of being worthy of respect, esteem, or honor. You may not know when you got it, but you sure as hell know when you don't.

I don't have to repeat all of the no-meal-break-no-pee-break-customer-using-the "F" word litany. You know it all too well. It is troubling that pharmacists have, as a group, allowed this to prevail. Can you imagine an attorney or an accountant tolerating what we put up with?

"Hang together or surely we will hang separately".

# Self-Respect

How hard is this one? When you are at the neighborhood summer picnic and in a group under the tree enjoying a cold brew with a lawyer, accountant, dentist and doctor and think you don't belong, trust me, you do not have self-respect.

Self-respect is simply a belief in your own worth and dignity. You have spent thousands of dollars and years becoming the medical professional who is the last word on drugs. How can you possibly feel so lousy about yourself? That is a rhetorical question for you to answer for yourself. When you have that figured out, you will be on your way.

When the modern pharmacy schools do not address this issue early on, they have graduates who are vulnerable to the present state of affairs. It is pathetic that new pharmacists are left on their own to navigate the perilous waters of retail pharmacy with no formal warning from their schools and even less guidance from their preceptors.

My criticism is aimed directly at an educational system that prepares students to be clinical mavens in a complicated seascape and then sends them out into an ocean that is teeming with sharks. How can the schools expect new pharmacists to swim in these waters with no warnings or preparation?

The easiest road to self-respect is to practice pharmacy exactly according to your own personal standards. If you feel really good about being a good manager of the Prescription Mill, then so be it. Who am I to comment about your satisfaction. However, if you need to do more. Do it!

# Integrity

*Say what you mean and mean what you say. Do what you say you are going to do and say only what you are going to do. Keep your word. Don't lie, especially to yourself.*

Most people don't realize that integrity is to help you, not to help others. Integrity is so important because often times it is the one thing inside of us that motivates us to keep going and stay strong. Having and using integrity means being like a rock; strong-willed, steady, and proud. Integrity is important because your good name is really the only thing you have. Once that is gone life gets a lot harder. If you have integrity, people will trust you and interact with you more easily.

# What will define Pharmacy in the 21$^{st}$ Century

It is easier to tell you what it will not be and that is the Prescription Mill. You will be too busy doing what a medical professional who is the expert on drugs should be doing. There will be a new designation of technician. Call the position "Specialized Technician" or "Advanced Technician". With specialized training there will be technicians checking the filling done by other technicians. The pharmacist will be called upon to make final choices in case of ambiguity or when the computer flags alerts.

You should welcome this. Pharmacists will use their training and experience to make more money for the company and, once this is noticed, you will get a piece of the action. If they don't want to share, go make money for another company.

I'll be happy if this little book causes a conversation to begin, everywhere. The most important place, of course, is at pharmacy schools. Among students and professors. Preceptors and interns. Also, in the offices of the companies that hire pharmacists.

Change, sea change is necessary for pharmacy and pharmacists to thrive in the medical professional landscape. I will enjoy watching and would love to be able to say, "Told you so."

*Come mother and fathers*
*Throughout the land*
*And don't criticize*
*What you can't understand*
*Your sons and daughters*
*Are beyond your command*
*Your old road is*
*Rapidly agin'*
*Please get out of the new one*
*If you can't lend your hand*
*For the times they are a-changin'*
*Bob Dylan*

Appendix

# What is Sexual Harassment?

Sexual harassment at work occurs whenever unwelcome conduct on the basis of gender affects a person's job, It is defined by the Equal Employment Opportunity Commission (EEOC) as unwelcome sexual advances, requests for sexual favors, and other verbal or physical conduct of a sexual nature when:

submission to the conduct is made either explicitly or implicitly a term or condition of an individual's employment, or submission to or rejection of the conduct by an individual is used as a ,basis for employment decisions affecting such individual, or the conduct has the purpose or effect of unreasonably interfering with an individual's work performance or creating an intimidating, hostile, or offensive working environment.

A second type of unlawful sexual harassment is referred to as hostile environment. Unlike a quid pro quo, which only a supervisor can impose, a hostile environment can result from the gender-based unwelcome conduct of supervisors, co-workers, customers, vendors, or anyone else with whom the victimized employee Interacts .on the job. The behaviors that have contributed to a hostile environment have included:

threats to impose a sexual quid pro quo.

discussing sexual activities;

telling off-color jokes;

unnecessary touching;

commenting on physical attributes;

displaying sexually suggestive pictures;

using demeaning or inappropriate terms, such as "Babe";

using indecent gestures;

sabotaging the victim's work;

engaging in hostile physical conduct;

granting job favors to those who participate in consensual sexual activity; using crude and offensive language

## When Does an Environment Become Sexually Hostile?

To create a sexually hostile environment, unwelcome conduct based on gender must meet two additional requirements: (1) it must be subjectively abusive to the person(s) affected, and (2) it must be objectively severe or pervasive enough to create a work environment, that a reasonable person would find abusive.

To determine whether behavior is severe or pervasive enough to create a hostile environment, the finder of fact (a court or jury) considers these factors:

The frequency of the unwelcome discriminatory conduct;

The severity of the conduct

Whether the conduct was physically threatening or humiliating, or a mere offensive utterance;

Whether the conduct unreasonably interfered with work performance;

The effect on the employee's psychological well-being; and

Whether the harasser was a superior in the organization.

## Is it Really Sexual Harassment?

Hostile environment cases are often difficult to recognize. The particular facts of each situation determine whether offensive conduct has "crossed the line" from simply boorish or childish behavior to unlawful gender discrimination. Some courts state that men and women, as a general rule have different levels of sensitivity -- conduct that does not offend most reasonable men might offend most reasonable women. In one study, two-thirds of the men surveyed said they would be flattered by a sexual approach in the workplace, while 15 percent would be insulted. The figures were reversed for the women responding. Differing levels of sensitivity have led some courts to adopt a "reasonable woman" standard for judging cases of sexual harassment. Under the standard, if a reasonable woman would fell harassed, harassment may have occurred even if a reasonable man might not see it that way.

Because the legal boundaries are so poorly marked, the best course of action would be to avoid all sexually charged conduct in the workplace. You should be aware that your conduct might be offensive to a co-worker and govern your behavior accordingly. If you're not absolutely sure that behavior is sexual harassment, ask yourself these questions:

Is this verbal or physical behavior of a sexual nature?

Is this conduct offensive to persons who witness it?

Is this behavior being initiated by only one of the parties who has power over the other?

Does the employee have to tolerate that type of conduct in order to keep his or her job?

Does the conduct make the employee's job unpleasant?

If the answer to these questions is "yes," put a stop to the conduct.

## How Can You Tell if Conduct is Unwelcome?

Only *unwelcome* conduct can be sexual harassment. Consensual dating, joking, and touching, for example, are not harassment if they are welcomed by the persons involved.

Conduct is *unwelcome* if the recipient did not initiate it and regards it as offensive. Some sexual advances ("come here Babe and give me some of that") are so crude and blatant that the advance itself shows its unwelcomeness. In a more typical case, however, the welcomeness of the conduct will depend on the recipient's reaction to it.

### Outright Rejection

The clearest case is when an employee tells a potential harasser that conduct is unwelcome and makes the employee uncomfortable. It is very difficult for a harasser to explain away offensive conduct by saying, "She said no, but I know that she really meant yes." A second-best approach is for the offended employee to consistently refuse to participate in the unwelcome conduct. A woman who shakes her head "no" and walks away when asked for a date has made her response clear.

### Soured Romance

Sexual relationships among employees often raise difficult issues as to whether continuing sexual advances are welcome. Employees have the right to end such relationships at any time without fear of retaliation on the job, so that conduct that once was welcome is now unwelcome. However, because of the previous relationship, it is important that the unwelcomeness of further sexual advances be made very clear.

### What Not To Do

Invite the alleged harasser to lunch or dinner or to parties after the supposedly offensive conduct occurred.

Flirt with the alleged harasser.

Wear sexually provocative clothing and use sexual mannerisms around the alleged harasser.

Participate with others in vulgar language and sexual horseplay in the workplace.

Even if you do not find the conduct personally offensive, remember that some of your co-workers might, and avoid behavior that is in any way demeaning on the basis of gender. In determining if your own conduct might be unwelcome, ask yourself these questions:

Would my behavior change if someone from my family was in the room?

Would I want someone from my family to be treated this way?

You and your employer share responsibility in maintaining a harassment-free work environment. Many organizations have written policies, distributed to all employees, that contain examples, that contain examples of prohibited conduct and describe procedures for handling complaints. These policies may forbid conduct that falls short of unlawful sexual harassment. It's important to learn about your own employer's policy.

Retaliation against any employee who reports sexual harassment or who cooperates when the employer investigates a claim of sexual harassment is prohibited. The employer will want to conduct a prompt and thorough investigation of all complaints, and matters will be kept as confidential as possible.

Employer policies typically provide that any employee found to have violated the policy will be subject to discipline, up to and including immediate discharge, and that the complaining employee will be told whether action has been taken, even if not told specifically what was done.

**Respond Appropriately When You Encounter Sexual Harassment**

If you experience sexual harassment or witness it, you should make a report to the appropriate official. You do not have to report the incident to your supervisor first, especially if that is the person doing the harassing. Before you report a problem, you might want to try some self-help techniques, using the DO's and DON'Ts listed below.

If you do follow these self-help suggestions, remember that sexual harassment is an organizational problem, and the employer wants to know about it so it can take prompt and appropriate action to ensure that no further incidents occur, with the present victim or other employees, in the future. Report incidents immediately, especially if they are recurring. Employees who promptly report harassing conduct can help their organization as well as themselves. One comprehensive survey by the American Management Association reported that roughly two-thirds of internal reports result in some kind of discipline being imposed on the alleged harasser, with even more internal reports resulting in either discipline or counseling.

## Keep It Confidential

First, whether you are the accused employee, the complaining one, or merely a potential witness, bear in mind that confidentiality is crucial. Two people have their reputations on the line, and you may or may not know all the facts. In the typical situation, the employer will keep the information it gathers as confidential as possible, consistent with state and federal laws, and both the accused and the complainant will have a chance to present their cases.

## Don't Be Afraid To Cooperate

There can be no retaliation against anyone for complaining about sexual harassment, for helping someone else complain, or for providing information regarding a complaint. The law protects employees who participate in any way in administrative complaints, and employee policies protect employees who honestly participate in in-house investigations. If you are afraid to cooperate, you should be very frank about your concerns when talking to the employer's investigator.

*Answer the questions completely.*

## As the Complainant

If you are making the complaint, the investigator will need to know all the details, unpleasant though they will be to recount. The investigator has a duty to be fair to everyone involved and needs as much information as possible. Be prepared to give the following information:

The names of everyone who might have seen or heard about the offensive conduct;

The names of everyone who may have had a similar experience with the alleged harasser;

A chronology -- when and where each incident occurred;

The reasons why you did not report the incidents earlier (if you have delayed at all)

Your thoughts on what the employer should do to correct the problem and maintain a harassment-free environment.

The investigator may need to talk with you several times while other employees are questioned and information is gathered.

## As the Accused  (As a Pharmacy Manager, you may be in a position of the accused)

If you are the person accused of sexual harassment, you must remember that you have a duty to cooperate in the investigation, regardless of whether you believe the allegations to be true or false. You will be expected to answer questions completely and honestly.

You may be asked not to communicate with certain individuals during the course of the investigation. You must remember that you are not to retaliate against the person who make the complaint or against anyone who participates in any way in the investigation. You must treat them in the same fair and even-handed manner you would if no complaint had ever been raised. Failure to abide by these rules may result in discipline against you, even if the investigation shows that no sexual harassment occurred. Indeed, retaliation against a complainant may violate the law even if the underlying complaint of harassment cannot be substantiated.

You should expect to be asked to confirm or deny each of the specific allegations against you. It is possible that the allegations are gross exaggerations or downright lies, but it is important to remain calm and keep your responses factual. You may be asked to provide any facts that might explain why the complainant would be motivated to exaggerate or fabricate the charges. The investigator might need to talk to you several times while other employees are questioned and information is gathered.

### As a Potential Witness

You may be asked to provide details concerning alleged sexual harassment between other employees. You have a duty to respond truthfully to the questions concerning these allegations. The natural tendency after an interview by an investigator is to share with co-workers the more interesting details. Remember that the employer's policy is to keep the interviews as confidential as possible. Gossip about allegations of sexual misconduct, can fairly damage the reputation of co-workers.

### Keep the Lines of Communication Open

The object of the employer's investigation is to find out what happened. The investigator may conclude that sexual harassment occurred, that it did not occur, or that it is impossible to tell what really happened.

As the complainant or as the accused, you have the right to know in general terms what the organizations conclusion is, and you should ask if you are not told. Do not assume that the matter is settled until you have been told so directly. If you are the complaining party, it is important to promptly report any new incidents of sexual harassment that occur after your first talk with the investigator, and to tell the investigator about anything you may have forgotten or overlooked. Do not be discouraged by the fact that the employer takes time to act, and bear in mind that the more information you provide, the better chance there is for decisive action by the employer.

If you are accused, do not be discouraged if the employer's investigation fails to completely clear your name. It is not uncommon to conclude that there is no way to tell what really happened. Remember, sexual harassment complaints often involve one-on-one situations where it is difficult to determine the truth. More over, two people can have totally different perceptions of the same incident. The best you can do in such a situation is to avoid further situations where your words or conduct can be used as evidence of sexual harassment.

**Expect Adequate Remedial Action**

If the employer finds that sexual harassment did occur (or even some inappropriate action falling short of sexual harassment), expect the employer to take some remedial action. A variety of disciplinary measures may be used, including:

An oral or written warning;

Deferral of a raise or promotion;

Demotion;

Suspension; or  Discharge

The action taken in any particular case is within the organizations discretion. The precise nature of the discipline is often kept confidential to ensure that the privacy of individuals is protected. One aim of the action is to deter any future acts of harassment. If you, as the complaining party, fell that the harasser is retaliating against you for complaining or continuing to harass you, you should immediately use the employer's procedures to report the conduct so that the employer can take further action as appropriate.

If the employer does not have enough evidence to reach a conclusion about harassment, it still might take other actions, such as separating the parties, holding training sessions on preventing sexual harassment, or having the affected employees certify that they have read again and fully understand the employer's policy against sexual harassment.

Note: Many organizations forbid conduct that falls short of unlawful sexual harassment and do impose discipline for conduct that comes to their attention as the result of a sexual harassment complaint, even if the conduct does not violate the law or the organizations harassment policy. For example, a manager who makes sexual advances to subordinates might be disciplined for exercising poor judgment, even if the sexual advances were welcomed; and an employee who engages in a single incident of offensive gender-based conduct might be disciplined for inappropriate conduct, even if the incident was not severe enough to create a hostile environment. The fact that an employer imposes discipline in response to a complaint of sexual harassment is not admission, therefore, that any unlawful harassment has occurred.

## The DO's and DON'Ts of Sexual Harassment

### Do

Admit that a problem exists

Tell the offender specifically what you find offensive

Tell the offender that his or her behavior is bothering you

Say specifically what you want or don't want to happen, such as "please call me by my name not Honey," or "please don't tell that kind of joke in front of me."

### Don't

Blame yourself for someone else's behavior, unless it truly is inoffensive

Choose to ignore the behavior, unless it is truly inoffensive

Try to handle any severe or recurring harassment problem by yourself - - get help.

### Understanding Sexual Harassment

After having read this, you should have a pretty good understanding of what sexual harassment is, how to prevent it, and what to do if you see it. For review and general guidance, here are some of the most commonly asked questions about sexual harassment. For more specific information, contact the human resources office.

Doesn't sexual harassment have to involve sexual advances or other conduct that is sexual in nature?

No. The 1980 EEOC Guidelines on Sexual Harassment do suggest that conduct constituting sexual harassment must be "conduct of a sexual nature," but it is just as wrong and just as unlawful to harass people with gender-based conduct of a nonsexual nature. Consider, for example, a man and a woman each holding the same kind of job in an organization. If their supervisor gives demeaning and inappropriate assignments (such as serving coffee, picking up dry cleaning, emptying a waste basket) to the woman, but not to the man, because of the woman's gender, that conduct, if sufficiently severe or pervasive, could amount to harassment on the basis of sex even thought the assignments are not sexual in nature but whether it was based on the victim's gender.

Can sexual harassment occur without physical touching or a threat to the employee's job?

Yes. The nature of sexual harassment may be purely verbal or visual (pornographic photos or graffiti on workplace walls, for example), and it does not have to involve any job loss. Any nonsexual but gender-based conduct that creates a work environment that a reasonable person would consider hostile may amount to sexual harassment.

I'm so mad at the person who harassed me and at my employer that I just want to sue. Should I even bother to complain under my employer's sexual harassment policy?

Yes. You owe it to your employer and to your co-workers to report through the organization's channels to give the employer a chance to solve the problem promptly, before others are affected. A prompt complaint is also something that you owe yourself, even if your sole concern is to sue your employer. If you fail to use internal procedures, the employer's defense team will be sure to use that fact to argue that (1) the conduct complained of never occurred, (2) the conduct was not really unwelcome, (3) the conduct was not sever or pervasive enough to create a hostile environment, or (4) the employer cannot be held responsible for preventing or correcting harassment that it did not know about.

Furthermore, under the 1998 decisions by the Supreme Court in Ellerth and Faragher, if the employer has an effective anti-harassment policy that the employee unreasonably fails to use, the employer may win the hostile environment lawsuit on that ground alone.

Failing to complain can be particularly harmful to your legal interests if you claim that harassment forced you to quit. It is hard to blame your employer for forcing you off the job if it could have corrected the conduct but was never given the opportunity to do so.

**Filing a Charge of Sexual Harassment**

If you believe that you have been discriminated against at work because of your race, color, religion, sex (including pregnancy), national origin, age (40 or older), disability or genetic information, you can file a **Charge of Discrimination**. All of the laws enforced by EEOC, except for the Equal Pay Act, require you to file a Charge of Discrimination before you can file a job discrimination lawsuit against your employer. In addition, an individual, organization, or agency may file a charge on behalf of another person in order to protect the aggrieved person's identity. There are time limits for filing a charge.

*Note*: Federal employees and job applicants have similar protections, but a different complaint process

If you file a charge, you may be asked to try to settle the dispute through mediation. Mediation is an informal and confidential way to resolve disputes with the help of a neutral mediator. If the case is not sent to mediation, or if mediation doesn't resolve the problem, the charge will be given to an investigator.

If an investigation finds no violation of the law, you will be given a Notice of Right to Sue. This notice gives you permission to file suit in a court of law. If a violation is found, we will attempt to reach a voluntary settlement with the employer. If we cannot reach a settlement, your case will be referred to our legal staff (or the Department of Justice in certain cases), who will decide whether or not the agency should file a lawsuit. If we decide not to file a lawsuit, we will give you a Notice of Right to Sue.

In some cases, if a charge appears to have little chance of success, or if it is something that EEOC doesn't have the authority to investigate, they may dismiss the charge without doing an investigation or offering mediation.

Many states and local jurisdictions have their own anti-discrimination laws, and agencies responsible for enforcing those laws (Fair Employment Practices Agencies, or FEPAs). If you file a charge with a FEPA, it will automatically be "dual-filed" with EEOC if federal laws apply. You do not need to file with both agencies.

**How to File a Charge of Employment Discrimination**

You may file a charge of employment discrimination at the EEOC office closest to where you live, or at any one of the EEOC's 53 field offices. Your charge, however, may be investigated at the EEOC office closest to where the discrimination occurred

Where the discrimination took place can determine how long you have to file a charge. The 180 calendar day filing deadline is extended to 300 calendar days if a state or local agency enforces a state **or** local law that prohibits employment discrimination on the same basis. The rules are slightly different for age discrimination charges. For age discrimination, the filing deadline is only extended to 300 days if there is a **state** law prohibiting age discrimination in employment and a state agency or authority enforcing that law. The deadline is **not** extended if **only** a local law prohibits age discrimination.

Many states and localities have agencies that enforce laws prohibiting employment discrimination. EEOC refers to these agencies as Fair Employment Practices Agencies (FEPAs). EEOC and some FEPAs have worksharing agreements in place to prevent the duplication of effort in charge processing. According to these agreements, if you file a charge with either EEOC or a FEPA, the charge also will be automatically filed with the other agency. This process, which is defined as dual filing, helps to protect charging party rights under both federal and state or local law.

## Online Assessment System

EEOC does not accept charges online. However, we do have an online assessment tool that can help you decide if EEOC is the correct agency to assist you. You can then complete an Intake Questionnaire that you may print and either bring or mail to the appropriate EEOC field office to begin the process of filing a charge.

## Filing in Person

Each field office has its own procedures for appointments or walk-ins. Please check the field office list for your office's procedures.

It is always helpful if you bring with you to the meeting any information or papers that will help us understand your case. For example, if you were fired because of your performance, you might bring with you the letter or notice telling you that you were fired and your performance evaluations. You might also bring with you the names of people who know about what happened and information about how to contact them.

## Document, document, document!

You can bring anyone you want to your meeting, especially if you need language assistance and know someone who can help. You can also bring your lawyer, although you don't have to hire a lawyer to file a charge. If you need special assistance during the meeting, like a sign language or foreign language interpreter, let us know ahead of time so we can arrange for someone to be there for you.

## By Telephone

Although EEOC does not take charges over the phone, you can get the process started over the phone.

You can call:

1-800-669-4000    to submit basic information about a possible charge, and we will forward the information to the EEOC field office in your area. Once the field office receives your information, they will contact you to talk to you about your situation.

## By Mail

You can file a charge by sending us a letter that includes the following information:

- Your name, address, and telephone number

- The name, address and telephone number of the employer (or employment agency or union) you want to file your charge against

- The number of employees employed there (if known)

- A short description of the events you believe were discriminatory (for example, you were fired, demoted, harassed)

- When the events took place

- Why you believe you were discriminated against (for example, because of your race, color, religion, sex (including pregnancy), national origin, age (40 or older), disability or genetic information)

- Your signature

Don't forget to sign your letter. If you don't sign it, EEOC cannot investigate it.

Your letter will be reviewed and if more information is needed, they will contact you to gather that information or you may be sent a follow up questionnaire. At a later date,EEOC will contact you and may put all of the information you sent us on an official EEOC charge form and ask you to sign it.


Resources You Might Find Interesting

http://www.jimplagakis.com
You can find a few years worth of Jim Plagakis' column JP at Large at the Drug Topics Website
http://www.drugtopics.com


You can also subscribe to have the digital edition of Drug Topics delivered directly to your in-box every month.
http://advanstar.replycentral.com/?PID=301&V=DIGI

JP at Large Up to Date is a compilation of the first 169 of Jim Plagakis' columns that first appeared in Drug Topics Magazine since January, 1989.  It is published by Advanstar and is available here:
http://www.industrymatter.com/jpatlarge.aspx

Contact Jim Plagakis at: jaypeegakis@gmail.com

www.ingramcontent.com/pod-product-compliance
Lightning Source LLC
Chambersburg PA
CBHW072346200326
41519CB00015B/3681